D1347594

Getting into
Engineering
Courses

James Barton and

hn Southworth

edition

trotman | **t**

Getting into guides

Getting into Art & Design Courses, 11th edition
Getting into Business & Economics Courses, 14th edition
Getting into Dental School, 11th edition
Getting into Engineering Courses, 6th edition
Getting into Law, 13th edition
Getting into Medical School: 2022 Entry, 26th edition
Getting into Oxford & Cambridge: 2022 Entry, 24th edition
Getting into Pharmacy and Pharmacology Courses, 2nd edition
Getting into Physiotherapy Courses, 10th edition
Getting into Psychology Courses, 13th edition
Getting into Veterinary School, 12th edition
How to Complete Your UCAS Application: 2022 Entry, 33rd edition

Getting into Engineering Courses

This sixth edition published in 2021 by Trotman Education, an imprint of Trotman Publishing Ltd, 21d Charles Street, Bath BA1 1HX

© Trotman Publishing Ltd 2021

Author: James Barton and John Southworth

1st–5th edns: James Burnett

British Library Cataloguing in Publication Data
A catalogue record of this book is available from the British Library.

ISBN: 978 1 912943 36 4

Printed and bound in the UK by Ashford Colour Press Ltd

Contents

About the authors

James Barton is Director, Recruitment at Mander Portman Woodward. As well as having previously authored *Getting into Medical School*, *Getting into Veterinary School* and *Getting into Physiotherapy Courses*, James sits on interview, medical and audition panels and has given careers advice, talks and public seminars across a range of fields, both domestically and internationally. This is his nineteenth publication for Trotman.

John Southworth is Principal at Mander Portman Woodward's London college. He has a BSc in Engineering and an MSc in Electronics and Guided Weapons, and has taught Systems and Control for nearly 20 years. He is particularly passionate about renewable energy and robotics. John is currently the Chair elect (2021) of the Independent Schools Association, and regularly supports students applying to engineering courses.

Acknowledgements

I wish to say thank you to all the previous contributors to this book, in particular Barry Sullivan and Dr Stepan Lucyszyn, whose words still carry so much wisdom. To Yuliya Kosko and EducAd Consultants for their support in the writing of this text. To all the student contributors who volunteered their personal statements: Maxwell Pierce, Will Bray, Will Arnold, Ally Denman and Mustafa El-Maghraby, as well as Gholam-hossein Malekmadani for his case study and Peter Hicks for his. And to everyone at MPW, a workplace that has given, and continues to give, so much to both those it educates and those who work there. It has been a privilege to have worked here for so long.

This book contains advice from a variety of professionals, institutions and those who are studying or working within the field of engineering. I have also scattered my opinions, based on over 20 years' experience of helping students with university applications, throughout the book and fully accept responsibility for any errors that appear.

James Barton
October 2020

This book is dedicated to the person who reminded me recently that if you risk nothing, you risk everything. After all, risk combined with necessity is the mother of all invention, and everything needs to be reinvented.

It is also dedicated to my sister, Gemma Barton. The kindest soul who only ever wanted to make a difference. Live for how you would want to be remembered.

Re-inventing the wheel: Introduction

Engineering is the closest thing to magic that exists in this world.

Elon Musk

To an extent Elon Musk is probably right; and we do have a lot to be grateful to engineering for, as it has influenced and changed our lives since the dawn of man and we need it to keep on doing so in order to adapt to survive in the future.

Engineering is the past, the present and the future. Ask anyone what are the greatest engineering ideas in history and you will get different answers. Stonehenge is arguably a good example to begin with, although modern engineering has not yet allowed a bypass to be built there yet. The Egyptians, the Greeks, the Incas and the Romans were all engineering innovators. The bicycle, computers, 3D printers – the characteristic feature of our world is that it will continue to adapt, and engineers are at the forefront of that.

Whether we are talking about new building projects, drones, 5G infrastructure, VR, roads, railways, Formula One on the cusp of changing to Formula E in the next 10 years, commercial flights to space, new super liners, the Hadron Collider, or just basic amenities, there is always something to be said about engineering. The profession is not often seen as having any glamour, and yet the exciting thing about a career in this sector is that it can be whatever you want it to be.

As you get the Southern railway line from Brighton into London, you pass through East Croydon, an area that is currently going through massive regeneration. As you arrive in to London, you pass Battersea Power Station, also in the throes of regeneration as it is turned into high-end flats and a shopping centre. As you pass over the bridge and look down the River Thames, what strikes you is the change that London has gone through throughout history. This is most seen through the engineering projects along the banks of the river: from the buildings dating back to Queen Victoria's Great Exhibition to the 21st-century Shard, the feats of structural and civil engineering are impressive, inspiring, diverse and challenging. In so many ways, design and engineering are intertwined; the responsibility of any architect and engineer is to ensure that those buildings fit within the legacy of the city.

Few things will happen in our lives without the influence of engineering in some respect or other yet the sector has never had as high a profile as, for instance, architecture, medicine or economics among the public. However, in recent years that has begun to change, as engineering projects ranging in scale from nanotechnology and the electronics that create your smartphone to amazing feats such as sending the Rosetta spacecraft to Comet 67P have captured the public's imagination. Engineers are also involved in environmental and safety issues, as well as helping to shape the world's future energy and raw materials needs.

Engineering projects are also being introduced to young children in schools: Barrow Engineering at the Royal Society of Engineers, for example, gives students in Barrow-on-Furness the chance to learn about engineering from engineers and technicians. Projects such as these are essential to help the profession attract new audiences, year on year.

If you are interested in engineering, you might be able to name some famous engineers – Thomas Edison, Isambard Kingdom Brunel, James Dyson, George Stephenson and Tim Berners-Lee, for example – but most engineers work as part of teams; teams that work on making our lives as fulfilling, safe and manageable as possible without people being aware of what they do. There would be no exciting new buildings or bridges without the structural engineers to work alongside the architects; no iPhones or iPads without the electronic, telecommunication and materials engineers; no websites without software engineers; and Rosetta would not have started its mission without the mechanical, chemical and aerospace engineers who created it and successfully launched it from Earth.

As you are reading this introduction, you are probably thinking about what opportunities exist for engineers. I hope that reading this book, coupled with your research into universities and careers within engineering, will show you that there is a myriad of exciting possibilities waiting for you within this field.

About this book

The aim of this book is to take you through the process of applying to study engineering, from choosing courses and universities through to postgraduate courses and career opportunities. It is set up like any good engineering project, and can be explained with a flow diagram (see opposite).

You can use this book as a source of advice and information by reading the chapters or sections that are of relevance to you. However, I recommend that you read the book from the beginning rather than dipping in and out of it, because you will then get a more complete picture.

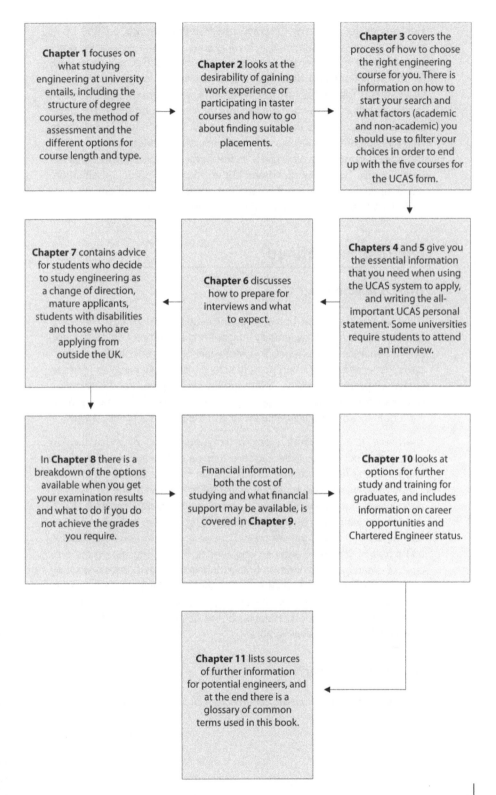

Chapter 1 focuses on what studying engineering at university entails, including the structure of degree courses, the method of assessment and the different options for course length and type.

Chapter 2 looks at the desirability of gaining work experience or participating in taster courses and how to go about finding suitable placements.

Chapter 3 covers the process of how to choose the right engineering course for you. There is information on how to start your search and what factors (academic and non-academic) you should use to filter your choices in order to end up with the five courses for the UCAS form.

Chapter 7 contains advice for students who decide to study engineering as a change of direction, mature applicants, students with disabilities and those who are applying from outside the UK.

Chapter 6 discusses how to prepare for interviews and what to expect.

Chapters 4 and **5** give you the essential information that you need when using the UCAS system to apply, and writing the all-important UCAS personal statement. Some universities require students to attend an interview.

In **Chapter 8** there is a breakdown of the options available when you get your examination results and what to do if you do not achieve the grades you require.

Financial information, both the cost of studying and what financial support may be available, is covered in **Chapter 9**.

Chapter 10 looks at options for further study and training for graduates, and includes information on career opportunities and Chartered Engineer status.

Chapter 11 lists sources of further information for potential engineers, and at the end there is a glossary of common terms used in this book.

References to university entrance requirements throughout the book are usually given in terms of A level grades, and the equivalent entry requirements for students studying the International Baccalaureate (IB), Scottish Highers and other qualifications can be found by using the UCAS Tariff (see Chapter 3). Universities, on their websites, provide further details of entrance requirements for all of the commonly accepted examinations, as does the 'Find a course' facility on the UCAS website (www.ucas.com). Regardless of the examination system you are using, the advice on applications given in the book is applicable to all candidates. If you have any questions about your own particular situation, universities are happy to deal with individual queries and can be contacted via their 'contact us' sections on their websites.

Engineering defined

What is engineering?

What is the difference between science and engineering? There are many definitions, but, essentially, engineering is the practical application of mathematics and science to create machines, processes or structures. Whereas the starting point in science generally involves trying to explain or predict phenomena through the development and verification of theories and models, engineering is the process of physically achieving a goal by applying scientific ideas and theories in a practical way.

There are many different types of engineering, all of which have a profound effect on our lives. Good examples of this are plentiful – from bridges to the redevelopment of cities, from Formula One to prosthetics – though a specific one for today would be the Elizabeth line on the London Underground network. This is scheduled to be operational in 2022, and has, as with most engineering projects, gone over budget. Adding a new tube line through the centre of London is an enormous undertaking and a feat of structural and civil engineering. From focusing on the process of creating the 21-km twin-bored tunnels through to the calculations to ensure no ground movement occurs that would impact on the structures above will all have combined physics, mathematics, design and construction. It is a good reflection of the holistic nature of the profession and the skills you will require.

What is an engineer?

Dreamer. Innovator. Researcher. Problem Solver. Inventor. Creator.

www.whatisengineering.com

What do engineers do?

Although we tend to classify engineers and engineering courses under different headings, there is a good deal of overlap, and most engineering projects or processes require the input of many different types of engineer. When you look in more detail at the course content of different engineering programmes, you will see that there are many common elements to these. For instance, mechanical engineers will spend time studying electronic and electrical engineering as machines often use electricity as a power source; and civil engineering requires an understanding of the properties of materials, as does structural engineering.

Most engineering projects involve the input of a range of specialisms (see Figures 1 and 2 below).

Figure 1 Input into building a skyscraper

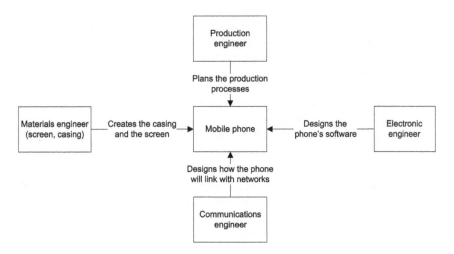

Figure 2 Input into building a mobile phone

Aerospace engineers

Aerospace engineering is concerned with the design of aircraft and spacecraft. It is made up of aeronautical engineering and astronautical engineering. The job of the engineer is concerned with the design and propulsion of the aircraft, as well as reviewing the effectiveness of the craft and the materials from which it has been constructed.

Agricultural engineers

Agricultural engineering focuses on everything from agricultural equipment and machinery through to bioenergy and development of resources on farms.

Biomedical engineers

Biomedical engineering (along with biotechnology) looks at the engineering aspects of living things, often for medical purposes. This can range from working with living materials, such as animal tissue, to the development of medical instrumentation (medical scanners and equipment used in surgery), and the design of machines and devices such as heart pacemakers and artificial limbs.

Chemical engineers

Chemical engineering deals with the industrial processes that produce, for example, drugs, food and fuels. Chemical engineers are concerned with not only the chemical properties of the materials they are producing or developing, but also the economic and safety aspects of the projects. Chemical engineering links closely with bioengineering, biomedical engineering and biotechnology.

Civil engineers

Civil engineering deals with the large-scale infrastructure that is an essential part of daily life, such as roads, bridges, dams, water supplies, and office and apartment blocks. (The name civil engineering came about in order to differentiate projects that were there to benefit society in general from military engineering projects.)

Design and product engineers

Design and product engineering deals with the process of creating and developing systems and devices. As well as looking at production processes, such as the design of production lines and factories, design and product engineers have to be aware of safety and cost issues. As an example, the production of a new car involves mechanical and electrical engineering input, computer hardware and software development, the choice of the right materials to ensure that the car will be safe, functional and attractive, and the creation of a production line that will be both efficient and cost-effective.

Electrical and electronic engineers

Electrical and electronic engineers work with electrical and electronic devices. As with mechanical engineering, these range from a microscopic scale (integrated circuits or solid state devices, for example) through to national electricity networks. Many universities offer a joint electrical/electronic engineering course, but it is also possible to specialise in just one of these subjects. There are close links between electronic engineering and computer or information technology (IT) engineering.

Energy engineers

Focusing on the energy industry, energy engineers work on electricity generation and transmission with an emphasis on alternative sources of energy, energy efficiency and environmental issues. Energy engineering is a rapidly growing field, encompassing traditional (fossil fuels, nuclear fission) and alternative (wind turbines, wave and tidal schemes, solar power, biofuels) sources of energy.

Environmental engineers

This field is solely focused on the protection and improvement of the environment and the world around us, including the removal of plastic waste, for instance.

Information systems engineers

Information systems engineering is closely linked to electronic engineering. It focuses on computer systems and the transfer of electronic and digital information – mobile phones, the internet and computer operating systems.

Marine engineers

Marine engineers look after anything at sea, including boats, oil rigs, ships, submarines, and so on. There is a need for engineering sciences and design principles in all aspects of shipping, drilling and transportation above and below the water, including vessels, pipes, rigging and cables and so on. With bigger and bigger cruise ships being built, more and more naval vessels being commissioned, and more work on wind farms at sea, this form of engineering will endure into the future.

Materials engineers

Materials engineering is closely related to structural engineering in that it looks at the properties of materials that are necessary to create structures. However, it also covers the use of materials for other requirements, such as plastics, ceramics, glass and polymers. A mobile phone manufacturer, for example, may work with materials engineers to ensure that the phone is strong enough to withstand daily knocks while at the same time being light and attractive.

Mechanical engineers

Mechanical engineers work in the development and manufacture of machines. This is obviously a very broad description: the word 'machine' covers an enormous range of devices, from medical equipment that is used to perform microsurgery through to aircraft carriers. Mechanical engineering courses include a number of specialist areas, such as aeronautical engineering and automotive engineering. Aeronautical engineers work on all aspects of aircraft and spacecraft, from aerodynamics to the design of engines. Automotive engineers can also work on things that fly, as well as cars and other forms of transport.

Petroleum engineers

Petroleum engineering covers all aspects of the oil, gas and petroleum industries – from exploration and excavation, through refining and purification, to distribution. It covers geological studies, the chemical properties of hydrocarbons, and industrial processes.

Process (systems) engineers

Chemical, mining, petroleum, biotechnology and pharmaceutical engineering can all be described as being branches of process engineering. Process engineers use chemical and biological processes to make raw materials into useful commodities. Process systems engineers tend to work on IT and computer programmes to facilitate these processes.

Sports engineers

This branch of engineering is concerned with the development, design and testing of sports equipment for the better of the game and the safety of the players; this is partly corporately driven.

Structural engineers

Structural engineering deals with the use and suitability of materials intended for creating structures such as buildings, bridges, stadiums and electricity pylons. The discipline covers the structure and properties of materials on a microscopic level, and the behaviour of structures on a macroscopic scale. Structural engineers work closely with architects and civil engineers. Designing a breathtaking structure based on an architect's sketch can be a creative and rewarding process, as can handing it over to the civil engineers to build. The career can also be attractive to those who like using modern computer aided design tools.

There are many other classifications of engineering disciplines and courses (see page 38), and there are subsets of some of the areas mentioned above and courses that combine two or more of these disciplines. So, you will need to spend some time researching possible courses before deciding what you want to do. Advice on this can be found in Chapter 3.

Opportunities for engineers

Very few careers provide as many opportunities as engineering while at the same time offering secure employment prospects.

- Engineers can work anywhere in the world.
- The work can be theoretical or practical, and can be carried out within an office or on-site.
- Engineers can work for multinational companies or set up their own business.
- Engineers can work on any scale, from nanotechnology and micro-electronics, through to building the world's biggest structures.
- The world will always need engineers.

Case studies

Case studies are an interesting and informative way to find out more about engineering. Most university websites will have case studies of graduates, and the professional engineering bodies (see Chapter 11) illustrate their particular disciplines using case studies. The Royal Academy of Engineering's website has a good selection and would make an ideal starting point for those researching a career in engineering: www.raeng.org.uk/education/this-is-engineering.

Case study

'I grew up watching the great, late Stirling Moss achieve what he did in the golden age of motor cars. It really inspired me and drove (forgive the pun) me to want to be part of that world. However, I was never really that interested in becoming a driver. I wanted to be responsible for what made these magnificent machines work.

'So I went and studied Mechanical Engineering at the University of Bath. It was an excellent course, but the path I wanted to take was not apparent to me at the time. Although a boyhood goal, motorsport at that point did not want me, or at the very least did not want to show me how to access it.

'Instead, I went to work as a mechanical engineer in the construction industry, which, though not what I had originally envisaged, was where I was really moulded in my early career. It gave me the opportunity to travel and to work on some exceptional projects. After doing this for about five years, I heard about an apprenticeship with Aston Martin. I applied and got offered a place, learning the job alongside some of the best car mechanics of their generation.

'I learnt a lot at Aston Martin, from the design process to the complexities of the engines. I was lucky enough to work on the development of the DB model and the V6 to V8 engines. I spent a very happy 12 years with the company before finally moving into my dream role. I remember meeting the Williams team boss at an event one day, and we got to discussing cars. From there, I received a call to interview for a role in the Formula 2 team, which I got and stayed with the team for eight years before moving into the Formula 1 paddock in a technical role.

'My story is my own, but if it comes to offering advice to any aspiring engineers out there I would say this: it is great to have dreams; however, never lose sight of the fact you might have to take a round-about route to get there. This is natural and means we are shaped by our experiences. I certainly became a better motorsport engineer from taking a more general mechanical engineering degree because my knowledge was broader.'

Maurice Byfield MEng

Working internationally

Few careers provide greater opportunities to work or study overseas, and this is one of the attractions of engineering. Many of the world's major engineering projects are undertaken by multinational companies, and many engineers work freelance, choosing their projects and locations to suit their skills and circumstances. The rapid technological advances of the BRICS nations (Brazil, Russia, India, China and South Africa) have created enormous demands for qualified engineers.

Sixth form considerations

Study combination

Generally speaking, it is considered that the majority of engineering courses will require mathematics and physics A levels. There are additional variables to this, in the cases of Biomedical Engineering and Chemical Engineering, for instance. The bigger question is what should the third subject be and what is the need for further mathematics? The best advice for the third subject is to take something that you are going to enjoy, as the grade will talk at the end of the day. That is not to say go off-piste in your choices, it is important to remember what it is that you are gaining from that qualification. A lot of people question the requirement for further mathematics at Sixth Form. The answer to that is ambiguous. It is not usually required by a university, unless overtly stated, because there are schools that don't offer it as a subject. That said, the wording can often imply that it would be preferred if you had it. The simple reading is that, if you can, you might wish to consider it because you will be studying further mathematics as part of an engineering course anyway; but it will not break your application if you do not have it.

T Levels

Launched in September 2020, a T level is a two-year qualification equivalent to three A levels. They have been developed alongside guidance from employers and businesses and aim to provide students with a platform from which they can develop skillsets for specific industries. The learning will be balanced between classroom teaching and on-site experience, for which students need to complete at least 315 hours of work (45 days). While not all subjects were ready to rollout in September 2020, it is to be phased in over three years.

There are two distinct engineering pathways:

- Engineering, processing, manufacturing and control (September 2022)
- Building services engineering for construction (September 2021).

For the grading and UCAS equivalent points, see Chapter 3.

EPQ

The EPQ enables students to develop and demonstrate particular interests outside of their core A level subjects. The EPQ is worth 50% of an A level (in terms of UCAS points) and can add value to Russell Group and/or Oxbridge applications.

'We welcome the Extended Project and would encourage applicants to undertake one as it will help to develop independent study and research skills valuable for higher education.'

University of Cambridge

Typical university offers to read BEng Engineering are AAA, AAB or ABB with an A*, and A at EPQ. The EPQ is not taught, rather students work independently to research and develop their project. In conjunction with their supervisor, the student chooses a topic and a format for the project. This can be a research-based essay (of between 4,000 and 5,000 words), a creative piece (such as a short play or short story), an artefact (such as a model) or a presentation.

Competition for places

The information below is subject to revision dependent on the Covid-19 guidelines, as mentioned below.

While the ratio of applicants to places for engineering courses is lower than for many other subjects, the competition for places at the higher-ranked universities is intense. Thus many candidates, while being successful at gaining a place on an engineering degree course, do so either through Clearing or at one of their lower-preference universities. You should aim as high as you can (within the boundaries set by your examination results and predictions) in terms of your choice of university, as employers will look not only at what you studied but where you studied. Therefore, do all that you can to ensure that your choice of university will stand you in good stead in the future.

In 2015, the government removed the cap on places available; however, this does not necessarily mean there are more places as the top universities are still constrained by space and laboratory/workshop facilities; and although the removal of the cap has given institutions more flexibility, there are still far more applicants than places at these institutions. In 2019, for engineering courses overall, there were 160,870 applicants, for 29,945 places, with the most popular areas being mechanical engineering (8,175), general engineering (5,115) and electrical and electronic engineering (4,485). Of the 29,945 successful applicants, 24,985 were men and 4,960 were women. As Table 1 below shows you, the competition for places at universities in the UK across all aspects of engineering is intense, and the picture is fairly constant year on year.

Table 1 Types of Engineering courses and application numbers 2018 and 2019

Type of engineering	2018 applications (number of places)	2019 applications (number of places)
General engineering	23,465 (4,845)	22,915 (5,115)
Civil engineering	22,710 (4,300)	22,535 (4,120)
Mechanical engineering	48,400 (8,245)	47,380 (8,175)
Aerospace engineering	20,345 (3,550)	21,580 (3,810)
Naval architecture	255 (70)	205 (50)
Electrical and electronic engineering	22,755 (4,535)	23,630 (4,485)
Production and manufacturing engineering	2,900 (550)	2,785 (615)
Chemical, processes and energy engineering	15,515 (2,910)	14,710 (2,590)
Other engineering	90 (10)	145 (35)
Combinations in engineering	5,130 (940)	4,985 (950)

Application ratios (the number of applications per place) vary from university to university and course to course. Most universities publish these on their websites. For example, the University of Bristol (www. bristol.ac.uk) had 8.8 applicants per place for engineering, University College London (UCL) (www.ucl.ac.uk) had 5 applicants per place for mechanical engineering and Manchester (www.manchester.ac.uk) had 6.4 applicants per place for chemical engineering.

As with most things in life, the more planning and preparation you do, the better your chances of success. This is, of course, true for your studies and examinations; but it is also the case that research and a well-planned UCAS application will give you a much better chance of being made offers by your chosen universities. This applies to all aspects of the application: choosing your courses and universities, looking at their entrance requirements, writing a personal statement that will demonstrate your seriousness about, and suitability for, the course, and ensuring that you are prepared for any interviews or entrance tests.

Ignore those who tell you that 'if you are lucky, you will get an offer' – if you pitch your applications at the level that is appropriate for your qualifications (past and predicted) and prepare properly, you can remove most of the uncertainty from your application. The following chapters will tell you how to do this.

The Covid-19 situation

At the time of writing, the world is experiencing a pandemic in the form of Covid-19. This has had an enormous implication on the schools and university markets, and is something to be aware of if you are applying in the next two years.

In spring and summer 2020, schools were closed and the final year examinations were replaced with moderated grades based on Centre Assessed Grades produced by the schools. This resulted in a political storm and a reversal to the original Centre Assessed Grades, and universities found themselves in the unenviable situation of having to accept all students with the revised offers, placing a premium on places not only for 2020 entry but also for 2021 entry. To ease overcrowding, some universities were even offering cash incentives to students to switch to 2021 entry, with deals extended to give remission from fees or accommodation.

Combined with the withdrawals of those students who attended university this year and deferred until next year due to Covid-19, means that a large proportion of university places have already been allocated before this year's UCAS cycle. The media would have you believe that it is a bleak picture therefore for students hoping to enter into university in

2021, though the picture may be a little more optimistic as we do not know if the cap on places will be lifted for a second year.

N.B. Covid-19 is likely to have an effect on work experience opportunities. However, there are online alternatives (see Chapter 2).

I was originally supposed to become an engineer but the thought of having to expend my creative energy on things that make practical everyday life even more refined, with a loathsome capital gain as the goal, was unbearable to me.

Albert Einstein, theoretical physicist

1 | The nuts and bolts: Studying engineering

What does studying engineering actually entail? Is it all practical work? How much mathematics is involved? Should I take a joint honours degree? This chapter covers these and many other questions. Once you have read the chapter, my advice is to go to the university websites as they will provide you with more detail as well as case histories and comments from current and past students.

Engineering courses

The structure of an engineering course will vary not only from discipline to discipline, but also from university to university. The common elements to all engineering courses are:

- mathematics (see page 20)
- physical laws (for example Newton's laws or the laws of energy conservation or thermodynamics)
- the physics of materials (physical, electrical and thermal properties)
- environmental, safety and health issues
- cost and other economic issues.

The difference between school and degree courses

'The students who succeed at university are usually distinguished more by their attitude to independent study, their passion for the subject and their willingness to get involved in other aspects of university life than they are by their entry grades. Engineering students work hard and play hard, and they frequently hold senior positions on student society committees or in volunteering groups.

'Staff often have to help students to see that bite-sized answers to questions that might have been appropriate at A level need to become well-argued and researched responses. Just because the question is only one sentence doesn't mean that the answer should be.

'A new development across the sector over the last 10 or 20 years has been the recognition that it is not enough to teach engineering students a multitude of technical skills. There is an increasing amount of discussion around professional responsibility, dealing with ethical dilemmas, learning from past disasters, etc. (see Chapter 6), which, if taught well, can be just as stimulating as the technical material. The Royal Academy of Engineering has been helping to pioneer these aspects and its website links to many useful publications in this area. So, a student who has studied mathematics, further mathematics and physics may have to adjust slightly as they realise that the broader skills they learned earlier in their school life will need to be dusted off and brought to bear.'

Lawrence Coates, Professor of Engineering,
University of East Anglia

Student experience

'When I was studying for my A levels, I knew what I wanted to do in the future, therefore my combination was easy. For all those considering whether to take further mathematics or not, I can tell you I wish that I had as the crash course at the start of the degree was tough. That said, the subject material was not and I gained a lot from working in small group sizes. University is the biggest eye opener you could ask for. Gone are the support structures from school – if you want help, you have to ask for it; however, the expectation is that you resolve things yourself. And that, at the end of the day, is one of the fundamentals of being an engineer. Having attended a sixth form college, I think I was perfectly ready to be a student and I had those learning skills ingrained into me. The advice I would give anyone is to make the most out of every opportunity and do not fall behind early, this is a degree unlike others in terms of workload.'

Mustafa El-Maghraby, Brunel University London,
BEng Automotive Engineering

Student experience

'I am just completing the fourth year of my BEng, during which I spent a year in industry. The course itself has been fascinating though exceptionally tough. There is a good variety in learning conditions and that really does inspire me. I am not someone to enjoy just sitting in lecture theatres, and that is not what the profession is about either. Lectures balance nicely with laboratory work and, of course, my industry placement, which has stood me in good stead as I have been offered a job at the end of this year.

'The grading of the course is a balance between written exams and lab reports and you are very much expected to work as part of a team, something I found very rewarding. That has been the huge appeal of the profession to me.

'Don't be surprised at how quickly the course accelerates beyond your A level knowledge; this can be daunting, but remember that everything worth having should never really be too easy to achieve as, this way, you gain a sense of accomplishment from what you solve.'

Laura Dehan, University of Sussex,
BEng Mechanical Engineering

Student experience

'I found the difference between A levels and university to be quite stark. At school, most of the learning takes place under supervised conditions and you are made to do lots of repetition and practice to help you learn. This is not the case in university because the teaching is done in lectures, so the reading, the practice and the learning generally take place without any supervision or direct guidance. The process of learning is much more independent at university because you are expected to know how to learn from your A level years and the staff have less time to answer questions and explain things outside lectures. At university there is often no immediate consequence for not completing work, but you must avoid letting yourself fall behind because you will have to do twice as much work to catch-up!'

Jordan Massiah, Magdalene College, University of Cambridge,
MEng Engineering

Look at the course content in detail on the university websites to ensure that your interests are covered. Two examples of the typical range of topics covered as part of a four-year (MEng) engineering degree are shown on the following pages.

Mechanical Engineering (Queen Mary University of London)

Year 1
- Mathematics and Computing for Engineers I
- Mechanics of Fluids I
- Engineering Mechanics – Statics
- Energy Conversion Systems
- Mathematics and Computing for Engineers II
- Engineering Design Methods
- Engineering Mechanics – Dynamics
- Thermo-Dynamics I

Year 2
- Grad, Div, Curl: Vector Calculus for Engineering
- Engineering Instrumentation
- Design for Manufacture
- Energy Conversion Analysis
- Solid Mechanics
- Control Systems Analysis and Design
- Engineering Materials for Design
- Heat Transfer and Fluid Mechanics I

Year 3
- Third Year Project
- Computer Aided Engineering for Solids and Fluids
- Heat Transfer and Fluid Mechanics II
- Materials Selection in Design
- Failure of Solids
- Environmental Engineering
- Combustion in Automotive Engines

Year 4

Range of options including:
- Computational Engineering
- Renewable Energy Sources
- Numerical Optimisation
- Computational Fluid Dynamics
- Robotics
- Vehicle Crashworthiness
- Advanced Gas Turbines

Electrical and Electronic Engineering with Management (Imperial College London)

Years 1 and 2
- Mathematics 1A
- Mathematics 1B
- Topics in Electrical Engineering
- Analysis and Design of Circuits
- Digital and Computer Architecture
- Programming for Engineers
- Electronics Design Project 1
- Mathematics 2
- Signals and Systems
- Electromagnetism
- Circuits and Systems
- Power Electronics and Power Systems
- Communications
- Control Systems
- Electronics Design Project 2

Year 3
- Accounting
- Entrepreneurship
- Embedded Systems
- Managing Engineering Projects
- I-Explore
- Individual Industrial Placement – Part 1 or Group Consultancy Project

Year 4
- Corporate Finance
- Managerial Economics
- Strategic Management
- Individual Industrial Placement – Part 2

Project work

Project work forms an important part of the later years of a degree course, and is arguably the most exciting part of the course. The differences between coursework/projects at school and at university are illustrated on the following page.

A level or equivalent	Degree level
● Your teacher helps you because he or she is judged on your results.	● Your lecturer provides you with a framework within which the coursework is completed.
● Study guides are consulted and marking guidance is utilised.	● Some constraints are explained in the handouts.
● Marks are maximised by ticking all the right boxes.	● You interpret them and add value or originality.
● Basically your teacher tells you what to do.	● Basically your lecturer never tells you what to do.
● Generally there is one right way to do it.	● Generally there are multiple right ways to do it.

Lawrence Coates, Professor of Engineering, University of East Anglia

The importance of mathematics

All engineers use mathematical methods as an integral part of their work. While much of engineering relies on physical processes to develop and produce devices, machines, structures, fuels or chemicals, these are all underpinned by mathematical calculations and models. If you look at the course outlines for engineering degrees you will notice that a significant amount of the first-year content involves mathematics. Most universities will specify an A level (or equivalent) mathematics grade in their entrance requirements, so if you are interested in becoming an engineer you will need to study mathematics. If you want to be involved in production or design but do not want to study mathematics, you could look at alternative courses such as product design.

Very few universities require students to have an A level in Further Mathematics or the equivalent in order to be considered for engineering courses because some schools don't offer it as an option. If they do not, they will generally be teaching you the equivalent during your first year of study. If you have an A level in Further Mathematics, and it is not a pre-requisite requirement, you are neither in an advantaged nor disadvantaged position in terms of application, though you are in an advantaged position once the course has started.

'Mathematics is essential and takes time to learn. A student who is competent in mathematics can probably learn any other engineering subject comparatively quickly. Frequently, when students ask partners from the engineering industry how much of the mathematics that they studied at university they actually use in practice, they are told "about 10%". This is a most misleading response. Mathematics trains students and graduates to think logically in everything that they do, including writing technical reports. This is one of the reasons why engineering graduates are sought after by so many different sectors of industry. So, although they may not use Laplace Transforms every day, the fact that they studied them in depth once will have developed their methodical approach to their work that avoids them making mistakes.

'Most universities provide additional mathematics support to engineering students. At the University of East Anglia (UEA) our Learning Enhancement Team employs experienced maths tutors as part of the learning support team, and they in turn co-ordinate a group of senior students who also offer help and guidance. So, a student who is prepared to work hard to develop their mathematics skills will usually find there is plenty of help available.'

Lawrence Coates, Professor of Engineering,
University of East Anglia

Methods of assessment and study

Most universities award degrees on the basis of examinations that are sat throughout the course, although the weightings between examinations sat in the different years of the course may vary from university to university. Most engineering courses also involve coursework, dissertations or practical assessments. Details are given on the university websites. Courses will also contain different amounts of laboratory or practical work, work placements with engineering companies, or on-site work experience. This will also be detailed in the course outlines on the university websites and so you can make sure that your chosen course suits your own preferences or requirements before applying. Teaching is normally conducted through lectures (sometimes supplemented by one-to-one tutorials) or laboratory workshops.

Some courses also have a work placement year as part of the course. For example, University of Sussex has either a four- or five-year Mechanical Engineering (with industrial placement year) BEng/MEng which, in Year 3, offers students the opportunity for a paid year in industry, which has been shown to help progress a graduate's career at the end of their course. Recent undergraduates have gone on to work for EDF Energy and GE Aviation. Other opportunities at other universities are available, and some even count the placement year as part of your assessment.

BEng vs MEng courses

BEng courses are generally three years in length, and the outcome is a bachelor's degree in the relevant area of engineering. MEng courses are generally four years in length and integrate the bachelor's degree with a master's degree. This is discussed in more detail in Chapter 10. Students who study MEng courses have a more direct route to Chartered Engineer status (see page 125) and will usually command a higher starting salary as well as gaining transferable skills such as teamwork and problem solving, because the master's component involves more group and project work. In most cases, students opting for the three-year BEng course have the opportunity to transfer on to the MEng course after Year 2.

Combined honours courses

There are a number of joint or combined honours degrees available, but there is less flexibility with engineering courses than there might be with arts or humanities subjects. This is simply because engineering degrees tend to lead towards careers in engineering, and so they focus on this.

Many engineers end up running their own engineering company or business, or taking a managerial role within an engineering firm, and so a wide range of degrees that combine engineering with management are available.

As an example, take the Electronic and Electrical Engineering with Management course offered by Imperial College. The programme for the first two years concentrates on the theoretical and technical aspects of electrical and electronic engineering. In the third and fourth years (there are four years because this is an MEng, not a BEng course, see page 18) there is more flexibility for you to steer the degree towards your own interests, and you would study business courses alongside the engineering courses.

If you intend to apply for a joint or combined honours course, you must ensure that your personal statement addresses both aspects of the course (see page 56).

Other examples of combined honours courses:

- Civil and Architectural Engineering – University of Bath
- Computer and Internet Engineering – University of Surrey
- Electronic and Electrical Engineering and Modern Languages – University of Sheffield
- Electronic Engineering and Computer Science – Aston University
- Electronic Engineering and Music – Bangor University
- Energy Engineering with Environmental Management – University of East Anglia

- Mechanical Engineering and Aerospace Engineering – University of Southampton
- Mechanical Engineering and Languages – Heriot-Watt University

Other courses

Foundation degrees

Foundation degrees are two-year full-time (or three-year part-time or distance learning) courses that are provided by some universities. (Do not confuse these with the foundation courses offered to some international students in place of A levels or the equivalent.) They are intended for students who do not have conventional academic back-grounds, for example students who left school after taking their GCSEs and have been working in a relevant field, or mature applicants. Many employers will accept Foundation degrees as an acceptable qualifica-tion; and there are many opportunities for students with a Foundation degree to follow this with an extra year of university study to gain a bachelor's degree. You can apply for Foundation degrees through UCAS (www.ucas.com). Subjects available at Foundation degree level include all the major engineering fields.

Degree apprenticeships

The vocational higher education market is growing in the UK, with more emphasis placed on teaching practical skills alongside qualifications, with a view to addressing the high-level skills gap that exists in certain industries. Degree apprenticeships are the older sibling of the T level (see page 11) and combines full-time paid employment with part-time study, enabling students to have the benefit of experiencing both work and study. This ideally suits anyone who is more practically minded and ultimately wants to start in the job market. The programmes will take from three to six years depending on the type of course you choose.

Engineering is typically an industry that lends itself well to degree apprenticeships and there are currently around 1,000 available across the spectrum of engineering qualifications, and the government aims to grow this number. There are currently over 60 universities implementing degree apprenticeships in the UK and the number of employers signing up to the scheme is growing. Roles vary from Materials Process Engineer to Rail and Rail Systems Engineer.

To see what options are available, visit www.ucas.com/alternatives/ apprenticeships/apprenticeships-england/what-apprenticeships-are-available/degree-apprenticeships or www.instituteforapprenticeships. org/apprenticeship-standards/?routes=Engineering-and-manufacturing &levelFrom=5&includeApprovedForDelivery=true.

Some examples of degree apprenticeships:

- Aerospace Engineering – BAE Systems
- Construction – E-ON Energy Solutions Ltd
- Nuclear Engineering – EDF Energy
- Engineering Degree Apprenticeship Programme – Rolls Royce

How much will a degree apprenticeship cost? Nothing. Employers are expected to cover the cost of apprenticeship training, and you will even be paid a wage while undertaking the apprenticeship scheme. This will, however, mean more competition for places. NB: You are not guaranteed a job at the end of a degree apprenticeship, but you will be employable within a skills-based sector, which will give you a significant advantage. Consult www.prospects.ac.uk/jobs-and-work-experience/apprenticeships/degree-apprenticeships, for more information.

Higher National Diplomas

Higher national diplomas (HNDs) are usually two-year courses, often equivalent to, or taught simultaneously with, the first two years of a bachelor's degree. Students who are successful on the HND course can study for a third year to gain a bachelor's degree. Entrance requirements are normally less stringent than for a degree. For example, Coventry University asks for BBC (112 UCAS Tariff points) from three A levels (including A level Mathematics and/or Physics) for entrance onto some degree-level courses, but only 40 points from two A levels, for the equivalent HND course.

Scottish degree courses

Undergraduate degree courses at Scottish universities leading to bachelor's degree qualifications are four years in duration, although it is sometimes possible to enter in year two. This is sometimes called 'Advanced Entry' and the advantage of this is that it reduces the course length to three years for a BEng course or four years for the MEng course. The Scottish universities that offer this option will specify the entry requirements for the second-year (level 2) entry on their websites. The grade or score requirements are generally higher than for the first-year (level 1) entry. For example, Strathclyde University asks for BBB at A level for first-year entry onto the Electronic and Electrical Engineering course, and AAB for entry into the second year.

The course structure of the four-year degree allows students to study a broader range of subjects in the first year, compared with what is on offer in the three-year degrees common in England, Wales and Northern Ireland. The MEng qualification (see page 18) normally takes five years in Scottish universities.

Case study

'I suppose I had decided on an engineering degree when I was in my early teens. At home, I was always practical and took things to pieces. I made structures out of Meccano and Lego, and learnt many wood-working skills using a variety of aged hand tools. I brewed homemade beer and wine, and developed a keen interest in viticulture. In the garden, I would build walls and concrete paths. I had an old computer and would take it to pieces and then rebuild it in a different way – so emerged my love of all things engineering, from civil to chemical, and from mechanical to electronic!

'At A level I studied Double Maths, Physics and Chemistry (although I found chemistry extremely challenging), and so my pathway to an engineering degree began. My memory of the first term was a huge emphasis on maths. Every day we had a maths lecture interspersed by sessions on Industrial Relations and some project and lab work. My strong advice to those going for an engineering degree is to ensure the university course includes lots of practical work, as this is by far the best way to learn engineering in my view.

'I was fortunate to then join the Army, and during that time had the opportunity to further my studies on a Masters course in Electronics and Guided Weapons. This further renewed my great love of electronics, as I studied Parallel Processing for Robotic Systems, something I then used when I joined the teaching profession to teach Systems and Control.

'Would I change anything? Not a chance! Life as an engineer has been incredibly fulfilling and has enabled me in later life to tackle many things, from car maintenance to house restoration, with confidence.'

John Southworth, MSc in Electronics and Guided Weapons

2 | Building blocks: Getting work experience

Work experience that is related in some way to engineering is an invaluable way of demonstrating to the universities that you are committed to the course. It also shows that you have researched how engineers translate their academic studies into the practical skills that are required in the real world. Work experience will also show you whether you are suitable for a career in engineering, and what qualities are needed in a successful engineer. One of the things that an admissions tutor will look for is how serious you are about your chosen course. By writing about your work experience and how what you saw relates to what you enjoy studying or to things that you have read about, you can show the selectors that you have thought about engineering as a whole and as a potential career, rather than as an academic discipline only.

Many surveys have highlighted the importance that engineering employers attach to internships and work experience. And this is also true of your university application. As more and more students chase a fixed number of places, preference is given to those candidates who can demonstrate that they have made an effort to find out what working within the field of engineering will be like.

Engineering affects every aspect of our lives, and engineers make things that need to be commercially viable, and so gaining work experience that is in some way related to engineering should not be difficult. If you are lucky enough to have some connection with an engineering firm then use this as a starting point. But there are many other part-time jobs and work placements that will allow you to investigate or experience aspects of engineering, such as:

- any company that physically makes something that is then sold commercially – light bulbs, toys, jam jars, window frames, car accessories
- farm work
- shops that sell or repair computers or mobile phones
- a local architecture practice
- a local builder.

What will you gain from work experience?

- You will have lots to write about in your UCAS personal statement, and you will be able to demonstrate your research into engineering.
- It will help you to decide which area of engineering is most suitable for you. Do you want to work outside or in an office? On practical problems or on the theoretical side of the subject? In a small business or a large multinational company?
- The contacts that you make during your work experience may be helpful in your future career.
- You will find out whether you would actually *enjoy* working as an engineer. Be aware of the quote from the American inventor and engineer Thomas Edison: 'I never did a day's work in my life. It was all fun.' If you don't think being an engineer would be fun then it is probably not the right career for you.

Looking for work experience

Where to start

You may be fortunate in that your school will arrange this for you as part of a work-experience scheme. If not, then you will need to look for placements yourself.

So, how do you get work experience?

- You could approach local companies or use any contacts that your family may have.
- The institutes of engineering have contacts with engineering companies so try to go through them (see contact details in Chapter 11). Some will also offer summer courses or 'taster' programmes. The Year in Industry programme provides help and advice for students aiming to use their gap year to gain work experience (www.etrust. org.uk/the-year-in-industry/about-yini).
- There are many schemes operated by universities and the engineering institutes aimed at attracting students into engineering, and these often involve work placements.

If you know someone, or of someone, in a local engineering firm, try asking to go in for one or two weeks' work experience or work-shadowing during the holidays. Remember that even a single day of work-shadowing is better than no evidence of experience within the workplace, and something even tangentially related to engineering is better than no work experience at all. Helping organise the files for a local car mechanic would give you some access to the practicalities of engineering problems, and would be a good stimulus for reading more about automotive engineering, for example.

Online work experience in respect of Covid-19

Through the pandemic, work experience has largely been cancelled. Nevertheless, while virtual work experience does not always lend itself well to engineering, there are companies, such as Jacobs (www.jacobs. com), who offer some level of training. Civil engineering firms are also offering four-week virtual training programmes, with modules around Computer Artificial Design (CAD) programmes.

NB: Try and avoid virtual work experience that have a fee attached – there is no need to pay extortionate sums to get what you can achieve yourself.

For information on the programmes available, look at SpringPod.co.uk, which gives links to a number of virtual courses; for example, the RAF, offers a virtual experience for engineers.

Students should also focus on making the most of YouTube to watch engineering videos. Also, it's well worth checking out Imperial College, london's Public Lecture series (www.imperial.ac.uk/computing/out-reach/public-lecture-series).

Summer and 'taster' courses

Another good way to learn more about engineering is to use part of your summer holiday on an organised programme that gives you a taste of what a career in engineering entails. A Headstart (www.etrust.org.uk) course is one of many options available and is highly recommended by university admissions tutors. The UCAS website's taster course search page is a good starting point: www.ucas.com/events/exploring-univer-sity/learn-about-uni-taster-course.

How to apply for work experience

You need to prepare a curriculum vitae (CV), sometimes called a résumé. This should be short and to the point, outlining your education, experience, achievements and contact details. It should be word-processed on plain paper (but can be accompanied by a handwritten letter), and it should not be longer than two sides.

Your CV should include:

- full name, address, telephone number and email address
- synopsis – two lines telling an employer what you want them to know about you
- education – places, qualifications and grades (start with the most recent)
- skills (e.g. computer skills, software packages you are familiar with, languages spoken, whether you hold a driving licence)
- work experience (full time or part time, with names and addresses of the companies or businesses and a brief description of your responsibilities)
- positions of responsibility
- hobbies
- names and contact details of two or three people who can act as your referees.

Points to remember:

- highlight any experiences or achievements that are relevant to engineering
- highlight any skills gained (teamwork, communication, responsibilities)
- ensure that the layout makes it easy to read
- check all spelling and grammar carefully
- make sure there are no gaps in the CV (periods of time that are unaccounted for).

A sample CV

Lay out your CV clearly and logically, avoiding gaps, and including any exams you are studying for as well as those taken. Below is an example.

George Edward Burrows

Address: Sharow Hall, Aylesford, Kent, ME20 MJB

Telephone: 0123 456 7890
Email: georgeeddie@btinternet.co.uk

Bright and enthusiastic student with a strong skillset in scientific research and innovation. Proven track record of people skills and teamwork.

Education
2011 – 2018: Wellington College
2018: A levels to be taken: Mathematics, Computer Science, Physics
2016: GCSEs: Mathematics (9), Geography (8), Dual Award Science (8,7), Electronics (7), Spanish (7), English (6), Psychology (6)

Work experience
Summer 2017
Mercedes Benz Service Centre
I repaired and serviced cars. I learned new procedures: how to top-up AC systems with Freon, how to adjust ignition timing and how to take apart, reassemble and change parts in both manual and automatic transmissions.
Summer 2016
Aston Martin Service Centre
I memorised the different models, the variations between them, and the details of spare parts. I learned wheel-alignment, how to apply diagnostics and to balance wheels with carbon ceramic brakes, as well as how to buff a car's paint-job. Most crucially, I learned the team-based aspect of this job, understanding that it is the sum of all parts that makes something work; I was commended for the role I played.

Skills
Modern languages: Spanish (intermediate).
IT: literate in all Microsoft programs and capable of coding in C++, Python and Java.

Positions of responsibility
Captain of 1st XV rugby team; prefect.

Interests
Sport, travelling and understanding other cultures and peoples, innovation.

References
Available on request.

The covering letter

The covering letter (often handwritten, unless the company asks for a word-processed letter) should add more detail to the CV and also explain why the post you are applying for is suitable for you. You should find out the name of the person who will read it, and refer to them by name. If you use the person's name, sign off with 'Yours sincerely'. If for some reason you need to address it to an unknown person, then use 'Dear Sir' or 'Dear Madam' and end it with 'Yours faithfully'. If you are asked to email your application, send the letter and CV as attachments and give a brief overview (name, why you are looking for work experience, brief details of current studies) in the body of the email.

A sample covering letter is shown below.

Mr P Cookson
Signet and Swan Automotive Engineers
Exeter
Devon

1 November 2020

Dear Mr Cookson

I am interested in applying for a temporary role or internship at Signet and Swan Automotive Engineers, to gain some experience before I apply to read Mechanical Engineering at university. I am especially interested in Signet and Swan Automotive Engineers as I admire a lot of the advancements you have made to high-performance cars in the past three decades, not least the development of the Caterham brand. This type of motorsport engineering has formed the basis of my EPQ , in which I am evaluating the future development of Formula-E balanced against the loss of classic cars.

In addition to my qualifications shown on the attached CV, I have been teaching maths part-time to school children, through which I have learned the importance of taking responsibility. At school, I am part of the Engineering Society, which invites speakers to talk to us once a month. Incidentally, Alfred Robin, one of your principal partners, was at the Goodwood Festival of Speed this year, and I had a good chat with him there about potential opportunities and advancements throughout the sector, and he agreed to do a talk in our school.

I hope that you will be able to consider me for a post. I can be contacted by telephone or email at your earliest convenience.

Yours sincerely

Julian Koskova

Work experience interviews

If you are lucky, you may then be called for an interview. Most of the advice given in Chapter 6, on university interviews, is also relevant to job interviews. Other points to remember are listed below.

- Do some research on the company in advance. Learn about what they produce and how long they have been operating.
- Ensure that you can explain how your particular skills and qualities would be useful to them.
- Be clear about when you can start, how long you can work for, and whether there are particular periods when you cannot work, for example on the day your examination results are released.
- Wear smart clothes, a business suit if possible. Ensure that you have polished your shoes and don't wear too much jewellery.
- Introduce yourself, address the main interviewer by name, and offer your hand for a handshake at the start and again at the end.

Student experience

'I found work experience incredibly useful as it helped me to confirm my interest in mechanical engineering and gave me an inside look into the day-to-day life of an engineer. Although I knew about what engineers did, I had never been able to experience it first-hand, and work experience gave me a unique insight into how projects come together and are completed. I found the knowledge that I gained from this very helpful when writing my personal statement and applying to universities, as well as useful for me in clarifying my future plans. I did placements at two different engineering firms and an innovation centre, all of which I organised by researching the companies and then reaching out via email to them. I enjoyed my work experience and would recommend that anyone wanting to study engineering undertake some.'

Holly McIntosh, UCAS applicant

How to use your work experience effectively

Undertaking work experience is a means to an end – to show the university selectors that you are a serious candidate. You will be expected to write about what you discovered about being an engineer in your personal statement, and to discuss it in more detail at your interview. It is useful to keep a diary of your experiences while you are there. Record what you did, saw and heard, and in particular conversations you had with engineers – not just technical things, but about their training and what they see as the main challenges and rewards in working in an engineering field. It is also a good idea to then do some research or reading on things that you encountered during your work placement.

> *A common mistake that people make when trying to design something completely foolproof is to underestimate the ingenuity of complete fools.*
>
> **Douglas Adams, author**

3 | Measure twice, cut once: Choosing your course

Choosing universities and engineering courses can be a bewildering experience because you will be confronted by an enormous number of options. This chapter covers the steps that you should take to narrow your search and to find the most appropriate courses for you, including what to look for in the course content of an engineering course, academic requirements, types of university and location. It also covers what the course or university can offer you in terms of your future career and what you will gain from it on a personal level.

What to consider

Academic considerations

You are allowed five choices on the UCAS application. The basic factors to consider when choosing your degree course are:

- the engineering course you are looking for
- where you want to study
- the kind of university you are aiming for
- your academic ability.

Non-academic considerations to bear in mind include:

- accommodation
- location
- sporting, musical or other extracurricular facilities
- costs.

You need to think seriously about your choice of universities, as the decisions you will take now may determine your future career options. At this stage, you may already have an idea of which universities you want to consider, based on the advice of friends and/or family, but you need to be as open-minded as possible. Make a list of between 10 and

20 universities in which you are interested; it is then important to reduce this to a much shorter list. Not only will the university be where you begin the next step in your education, it will also be your home for three or four years, so think carefully about the location, environment and accommodation options as well as the suitability of the course(s) on offer.

Here are some things to research and consider for each university.

- Get hold of the prospectuses and any departmental brochures for more details. Remember that university publications are there to attract applicants as well as to provide information, and may be selective about the information they provide, so read all of it bearing this in mind.
- Visit the websites of the universities you are considering. This is the best place to look for the most current information about a university. Another useful element to university websites is information on past and present students from a range of disciplines who give their views on student life at the institution. Some university websites have email links to current students who can answer any questions that you may have.
- Find out when the open days are and go to them if you possibly can. You will have a chance to look at the facilities, talk to current students and find out more about the course. Currently, due to the Covid-19 pandemic, and possibly looking ahead to the next couple of years, virtual open days have had to be more prevalent. Check university websites for when these are happening as they provide a valuable insight into university life.
- Discuss engineering and studying it as a subject with people you know who work as, or with, engineers; ask for their views on the reputations of different universities and courses. This may bring up some highly rated engineering institutions you may not have thought of.
- Investigate the grade requirements and be realistic about the grades you are expecting – your teachers at school or college will be able to advise you on this.
- Check that the course allows you to choose the particular options in which you are interested. If you are considering, for example, mechanical engineering but have a particular interest in aeronautics or automotive engineering, make sure that these options are available. You will not always know what each option actually covers by its title, so read the department's own prospectus carefully and address any unanswered questions by contacting the admissions tutors directly – contact details are usually on the departmental website.
- Think about whether you would like your course to include a placement with an engineering company. This might be for a few weeks, a term or even a year. If this is something that interests you, find out who organises the placement – you or the university – and whether there is any funding to cover, for example, travelling costs or subsistence.

- Consider whether you want to spend some time abroad. Many engineering courses offer the option of a year abroad, studying at a partner university. If so, think about practical details such as language or visa requirements, and costs.
- Investigate how much practical/ laboratory work is included in the course, and what are the practical engineering facilities, e.g. whether the facilities include state-of-the-art machinery or testing equipment.
- Look at what IT facilities are offered. If you do not have your own laptop, will the university have facilities for you to manage without one of your own? Access to computing facilities can be very important when you are working on a dissertation. Find out whether there is WiFi access in the study areas and/or accommodation.
- Ask about the reading lists, and whether the books are available in the library. Are you required to buy your own textbooks? If so, are second-hand copies available?
- Look up the specialties of the engineering course staff. Are they experts in the field of engineering in which you are particularly interested? Use the internet to find out what their experience is and what they have published, as this will give you a better indication of a department's strengths.

Student experience

'I needed a lot of university guidance to understand where the best place was for me to study at university. There are so many fantastic engineering degrees. My best advice to anyone is to, if possible, visit the campuses and ignore league tables – choosing a university is so much more than just the course, it is about whether the university offers the right experience for you. And I did not get it wrong.'

Joey Relisan, University of Southampton, BEng Electronic and Electrical Engineering

---

League tables

Newspapers often feature university rankings or league tables but there is no official ranking of universities or university courses in the UK, and so these tables are created using criteria selected by the newspapers themselves. There is a significant amount of variation between these tables, because each table will score the universities in a different way.

However, as long as you approach these rankings with caution, they can be a useful aid to the selection process, particularly if you look at how the rankings are assessed rather than simply looking at a university's position in the tables. There will be some criteria that you might regard as being important to you – graduate job prospects, for example – while you might not be so interested in the student to teacher ratios.

A good starting point is the *Guardian* newspaper's university rankings. The general rankings can be found at: www.theguardian.com/education and there is a link to the subject-specific rankings which feature a number of engineering disciplines. Of particular interest might be the scores based on job prospects, and the student satisfaction with the course and with the teaching.

You might be interested in how the UK's universities are regarded on the global stage, particularly if you are an international student or you are planning on working abroad at some stage in your career. There are a number of world university rankings which you might find useful. But bear in mind that, like the UK university rankings, they are not in any way 'official' rankings and are based on criteria that the organisations compiling them see as being important.

One example is the world university rankings compiled by *Times Higher Education* (www.timeshighereducation.com).

There are many other league tables and rankings, for example the one produced by *The Sunday Times* newspaper for UK universities, or by QS, which specialises in business education, for world rankings. No two rankings will produce the same results and so you need to use them as part of the process in making your decisions, not as the sole reason for a choice of university.

Another useful reference point is the National Student Survey, which offers a student perspective on individual degree courses at any given university. The questionnaire is completed by graduates across the UK; the results are published annually and can be viewed online at www.officeforstudents.org.uk/advice-and-guidance/student-information -and-data/national-student-survey-nss.

After completing your research, you should be able to narrow down your original list to the five choices for your UCAS form. Once you have done this, discuss the list with your teachers to see whether they think it includes sensible choices. They may ask you to think again about some of the choices.

Common areas of concern for a school's UCAS adviser are:

- your current subjects don't fit the requirements of the courses you are applying for
- that the grade requirements are too high (or too low) for the applicant's likely academic achievements
- that there is too much variation within the choice of courses to enable the applicant to write a coherent and focused personal statement (see Chapter 5).

Choosing the right course

Most universities offer a wide range of engineering courses, and it is important to investigate these thoroughly before making a choice. A particular university might list the following on the UCAS website:

- Aeronautical engineering
- Aeronautical engineering with a year abroad
- Biomedical engineering
- Biomedical engineering with a year abroad
- Chemical engineering
- Chemical engineering with a year abroad
- Chemical with nuclear engineering
- Civil engineering
- Civil engineering with a year abroad
- Electrical and electronic engineering
- Electrical and electronic engineering with a year abroad
- Electrical and electronic engineering with management
- Information systems engineering
- Materials science and engineering
- Mechanical engineering
- Mechanical engineering with a year abroad
- Structural engineering

University websites and prospectuses can provide more details on specialisms, and it is well worth spending time going through these. Some universities offer integrated or general engineering courses that allow students to choose their specialisation later in the course. If you are truly undecided about which particular area of engineering is the most suitable for you then this might be a good option. However, even if you apply for an integrated or general engineering course, it would be a good idea to have an idea of which specialisations interest you most as this will probably be asked at the interview.

When considering possible courses, read through all of the course content. Do not choose a course just because of its title. Courses with the same name at different universities can vary immensely in their content, and within the courses themselves the likelihood is that you will have a range of options to choose from once you start your course. This is also important if you are interviewed (see Chapter 6) because you may be asked to justify your choice. Being able to discuss the course structure in detail will be an important factor in convincing the interviewer that you are a serious applicant.

Similar-sounding courses also do not always have the same entrance requirements (examination results and preferred A level subjects). Examination results are specified as either grade requirements (for example, AAB) or Tariff points (for example, 136 – see the section on the UCAS Tariff below). Unless you are applying post-results (as a mature applicant or during your gap year), your referee will be asked to predict the grades that you are expected to achieve in your examinations. You should find out in advance what he or she is going to predict, because this will determine your choice of universities and courses. For example, if you apply for five university courses that require AAB but your A level predictions are BBB, there will be a high chance of being rejected by all of your choices. You will then have to try to find alternatives through the UCAS Extra scheme, or through Clearing (see Chapter 8). Similarly, if you are predicted to achieve A*AA, you are probably aiming too low if all of the courses you are applying for require CCC at A level.

As a rough guide, if you are predicted, say, ABB, it would be risky to choose a course that requires AAB. It would be safer to choose three or four that require ABB, and one or two that require BBB. This not only means that you have a good chance of getting a number of offers, but it also gives you options if you do not quite meet the grade requirements (see Chapter 8).

The UCAS Tariff

While some universities make offers based on grades at A level, Scottish Highers or IB scores, others make offers using the UCAS Tariff. The UCAS Tariff is a points-based system that compares different education systems or achievements. For example, ABB at A level is equivalent to 128 Tariff points, roughly equivalent to an IB Diploma score of 27.

UCAS Tariff for A levels
A* grade – 56 points
A grade – 48 points
B grade – 40 points
C grade – 32 points
D grade – 24 points
E grade – 16 points

AS scores are now worth 40% (or as close as possible) of the full A level UCAS Tariff points at each grade. D* in the BTEC and H7 in IB is valued at 56 points, and a top grade in a Scottish Higher is 33 points. An EPQ is 50% of an A level, and an A* in the EPQ is worth 28 points.

UCAS Tariff for T levels
AAA* grade – 168 points
AAA grade – 144 points
BBB grade – 120 points
CCC grade – 96 points
DDD grade – 72 points

Full details can be found on the UCAS website at: www.ucas.com/ucas/tariff-calculator.

Sandwich courses

Sandwich courses, either BEng or MEng, offer students the chance to spend some time gaining industrial experience through work placements organised or suggested by the university. Typically, this might involve taking a year away from the university, although some universities will also offer a variety of shorter placements. Students who choose sandwich courses do so because:

- they want to gain experience of working to enhance their future job prospects
- they want to learn more about what working as an engineer entails
- they want to make contacts in the engineering world to help them to find jobs in the future.

Overseas study

Studying abroad could also be a factor that affects your degree selection. It is possible to study engineering in many countries as part of a degree based at a British university. Not all of these courses send you off for a full year, though: there are schemes that last for only one term or semester. You do not need to be a linguist either, as it is always possible to study overseas in an English-speaking location such as North America, South Africa, Australia or Malaysia. As an example, if you study Electrical and Mechanical Engineering with International Study at the University of Strathclyde you will have the opportunity to study at a partner university in Year 4. The university's partner universities include institutions in Canada, USA, China, Singapore, New Zealand and Australia.

The availability of student exchanges has increased through programmes such as Erasmus+, which encourage universities to provide international opportunities where practical – particularly in Europe. The popularity of overseas study has encouraged some universities to develop special exchange relationships with universities further afield.

As a result of Britain leaving the European Union in January 2021, the Erasmus+ programme will be replaced with the government's new Turing scheme. The UK government claims it 'remains committed to international education exchanges, both with the EU and beyond'. For further information please see https://www.gov.uk/government/news/new-turing-scheme-to-support-thousands-of-students-to-study-and-work-abroad.

Many universities will provide contact details or testimonials from students who have studied abroad as part of their degrees. For example: www.strath.ac.uk/studywithus/studyabroad/goingabroad/international exchange/studenttestimonials.

Academic and career-related factors

Academic ability

It is important to be honest about what you think you will achieve in your A levels or equivalent because, for most, this is the deciding factor for selection. The best way to get a strong sense of your predicted results is to speak to your teachers.

Remember, be realistic; you may think that you can do much better than your teachers' predictions or than your AS (either internally or externally assessed) results indicate, but the predicted grades and previous results will go on your UCAS application form, and so it is important to ensure that the courses you apply for are consistent with your likely results. If you do much better than expected, you can always 'upgrade' using the UCAS Adjustment scheme (see page 115). Check also whether the universities are likely to make a grade or Tariff points offer. If the grade requirement is, for example, AAB, don't assume that a combination of grades that gives you the same Tariff points (136, achieved by gaining A*BB) would satisfy the university's requirements.

For more details about UCAS and filling in your application, see *How to Complete Your UCAS Application: 2022 Entry* (Trotman Education).

Educational facilities

Take a look at the facilities the university has to offer. Here is a checklist of what to look out for:

- access to lecturers if you need help
- computer facilities
- course materials
- laboratory/workshop provision
- lecture theatres
- library facilities
- multimedia facilities
- study facilities.

Quality of teaching

The Office for Students, the Higher Education Funding Council for Wales, the Scottish Funding Council and the Department for the Economy assess the level of teaching across the UK. Their findings are publicly available – see www.officeforstudents.org.uk, www.hefcw.ac.uk, www.sfc.ac.uk and www.economy-ni.gov.uk. League tables (see page 36) normally incorporate these into their rankings. From September 2017, the government introduced the Teaching Excellence Framework (TEF), which is there to assess universities and colleges on the quality of their teaching and is intended to help students with their university choices. It is worth noting this is a voluntary scheme and universities can receive a bronze, silver or gold award reflecting their excellence in: teaching, learning environment and student outcomes. Without a TEF award, universities cannot charge a student more than £9,000 per year.

Type of institution

There are four types of institution from which you can obtain a degree:

- 'old' universities
- 'new' universities
- private universities
- higher education colleges.

The 'old' universities

These are traditionally seen as the most ancient universities (such as Oxford, Cambridge, Edinburgh, University College London, Imperial College and Durham), along with those established in the early and mid-20th century (such as Bristol, Leeds, Liverpool and Reading), plus those established in the 1960s (such as Sussex, Bath and Stirling); usually with higher admission requirements, and with a strong emphasis on research.

The 'new' universities

Pre-1992 these were polytechnics, institutes or colleges, for example Kingston, Central Lancashire and Westminster. These tended to focus more on vocational courses with less emphasis (or none at all) on research. There are a number of excellent engineering degree courses at new universities that are very well regarded and highly competitive to get into, and they are often more flexible in terms of sandwich courses, work placements or links with industry.

The private universities

New private universities continue to join the university landscape and can offer well-funded departments, though with a high price tag. At this stage, there are no private universities offering undergraduate bachelor's degrees, though Arden University offers an MSc in Engineering Management. This may change over time.

Colleges of higher education

These are specialist institutions that have links with universities. The university awards the degree and can deliver part of the course, along with the institution. Many colleges also offer pathway programmes, such as Access or foundation courses for students who do not have the necessary academic qualifications for direct entry to a university.

Non-academic considerations

The starting point for your research ought to be the type of course you want to follow. Once you have done this you should then look at the non-academic facets of the universities that you are considering. These are irrelevant to the course, but instead concern what the university can offer you on a personal, social and cultural level.

Finances

Finance is an important factor to consider, as you will need to juggle a lot of outgoings when you go to university. You will need to take into account:

- accommodation costs
- availability of part-time work in the area to earn some extra money
- living costs, such as food
- 'lifestyle' costs
- travel from your accommodation to the university during term time
- travel from your home to and from the university for holidays
- proximity to your home, family and friends – will it cost you a lot of money to visit friends or to go home?

Accommodation

The university is likely to be your home for three or four years, so think carefully about the location, environment and accommodation options. Accommodation can vary wildly between institutions so you will need to think about where you would feel most comfortable. Do you want to live in halls of residence with other students, or independently (or with friends) in rented houses or flats? Do you want to be near your lectures or are you happy to live further away from the university?

Universities all have accommodation offices that provide help or information for students who do not want to, or are unable to, live in university-provided accommodation. Most universities offer arranged accommodation for first-year students in halls of residence, which may either be owned by the university or be shared with other institutions, but often students are expected to find their own accommodation in subsequent years.

At the time of writing, students needed to consider whether they wished to stay on campus given that universities were largely delivering online teaching during the Covid-19 pandemic.

Entertainment

You will be spending the next few years in a new place so you will need to have a look at the entertainment facilities it has to offer both within and outside the university. Are your particular interests or hobbies catered for? If you're a keen sports enthusiast, have a look at the facilities on offer and the sorts of clubs and teams you can join.

Site and size

- Campus university outside a town or city?
- Campus within a town or city?
- University buildings at various locations within a town or city?
- Large or small?

While some students have a clear picture of where they want to study, others are fairly geographically mobile, preferring instead to concentrate on choosing the right degree course and see where they end up. But university life is not going to be solely about academic study. It is truly a growing experience – educationally, socially and culturally – and so you need to do your research and think carefully to ensure that you choose the best university and course for you.

What can the university offer you?

This is possibly the most important question for you to consider. Universities need students. Even the most popular and oversubscribed universities are constantly looking to attract the best students. League tables and advice from teachers, employers or friends can help you in choosing the most suitable courses, but if you plan your application carefully you will have some choices to make. In this chapter we have looked at a range of selection criteria: location, accommodation, the course structure, facilities and other things. You might also want to investigate whether scholarships are available (see Chapter 9) and also the feedback from students who have previously studied at the university. The *Guardian* newspaper's university rankings incorporate scores based on student satisfaction with the course and with the teaching. 'Which? University' (www.uniguide.co.uk) also gives statistics on drop-out rates, ratios of male:female students and other information that you might find useful.

> *Science is about knowing; engineering is about doing.*
>
> **Henry Petroski, Professor of Civil Engineering**

4 | Righty tighty, lefty loosey: Completing your UCAS application

This chapter is designed to help you complete your UCAS application. Further advice on filling in your application is given in *How to Complete Your UCAS Application*, which is updated every year.

The UCAS application

The UCAS application is completed using Apply, the online application system on the UCAS website. There are five sections to complete.

1. **Personal information:** name, contact details, home address and nationality (for fee purposes – see Chapter 9).
2. **Your choices of university and course:** you have up to five choices of university or courses. This is entered by using codes for the university and the course. You need to take great care in ensuring that this is accurate. You will also be able to specify the year of entry (are you planning on a gap year?) and whether you want to live at home. Some universities, such as Oxford, Cambridge and Durham, operate a college system so you also have to choose the college at which you wish to be based.
3. **Education:** past examination results, details of your previous and current schools and colleges, and future examinations. This section requires particular care, and you will need to discuss this with your school or college to ensure that, for example, AS results are correctly entered if you sat exams externally rather than as an internal assessment (see Chapter 3).
4. **Employment:** have you been in full-time employment?
5. **Personal statement:** see Chapter 5.

When you have completed the application, you save it and mark it as complete. You then send it to your referee via the UCAS website. He or she will then complete the reference and, if applicable, add predicted grades or scores. If you are applying as a private candidate, you will give the contact details of a referee and the university will contact him/her directly (see Chapter 7).

University engineering departments are looking for motivated, well-qualified individuals, and so they are keen to provide as much encouragement and practical advice as possible. If you look hard enough, you will find lots of information from the universities that will help you with your application.

All universities provide detailed information on their websites about their selection criteria, including advice on writing the personal statement and an indication of how they assess each candidate.

Suggested timescale

Use the timescale below to help you plan your application.

Year 12

In year 12, during your first year of A levels or IB, you should start to think seriously about what type of engineering course you want to follow. Talk to as many people as possible – your teachers, family and careers advisers. Don't just focus on what area of engineering interests you most, but also on whether you want a three-year BEng course or a four-year MEng course, whether you want a campus university or one in the centre of a city, close to home or in another part of the country.

Your next step should be to find out when the universities that interest you have open days and arrange to visit them. After this, you ought to be in a position to make up a shortlist of courses and universities, ready for your application. You can start to order prospectuses from the universities, or download the PDF versions from the university websites. Remember to make a note of the grade requirements for your chosen universities.

In August, refine your choices in the light of any AS results or predicted examination results.

Year 13

In September of year 13, you can submit your UCAS application. The deadline for applications is 15 January, but I recommend that you apply as early as possible. If you are applying to Oxford or Cambridge universities, your application has to be submitted by 15 October. Some universities will want to interview you, and this can happen from November onwards. If you are required to sit entrance tests, you may be asked to do this in the first week of November.

You should start to receive decisions from the universities from January onwards, although some universities might get back to you earlier than this. You can keep up to date with the status of your applications using the online UCAS Track facility. You will receive one of three possible responses from each university:

- conditional offer
- unconditional offer
- rejection.

If you receive a conditional offer, you will be told what you need to achieve in your A levels. This could be in grade terms, for example AAB (and the university might specify a particular grade in a particular subject – AAB, with an A in mathematics), or in UCAS Tariff points (136 points from three A levels – see Chapter 3).

Unconditional offers can be given to students who have already sat their A levels, such as gap year students applying post-results. In recent years, 'conditional unconditional' offers have been used to force students into making a decision early on their university choices. However, the universities regulator has banned these offers until at least September 2021 amid concerns it was undermining the integrity of the university system.

Rejection means that you have been unsuccessful in your application to that university. If you are unlucky enough to receive rejections from all of your choices, or you decide to withdraw from your choices, you can then use the UCAS Extra scheme from February to apply to other universities, or to add choices if you did not use all of your five choices in the initial application. Extra allows you to add one new choice at a time. If you are successful in gaining an offer from your first Extra choice and you accept it, you are committed to accepting the place; if you decline the offer or are rejected then you can approach another institution through Extra. According to UCAS, around 180 students gained places on engineering courses through the Extra system in the 2019 application cycle.

Once you have received responses from all five universities (or offers through UCAS Extra), you will be given a deadline by UCAS by which time you have to choose one university as your firm acceptance, and a second insurance offer, normally one that requires lower grades. Once you have accepted your first choice university, you will be sent information about accommodation and fees and other practical information directly from the universities.

You sit your examinations in the summer, and receive your results in July or August:

- A level results are published in the third week of August
- Scottish Higher results are released in the first week of August
- IB results come out in the first week of July.

When the exam results are published, UCAS will get in touch and tell you whether your chosen universities have confirmed your conditional offers. Do not be too disappointed if you have not got into your chosen institution; just get in touch with your school/college or careers office and wait until Clearing begins in early July, when all remaining places are filled. You will be sent instructions on Clearing automatically, but it is up to you to get hold of the published lists of available places and to contact the universities directly. UCAS reports that around 3,805 students gained engineering places through Main Scheme Clearing (where an applicant was unsuccessful in the main scheme, i.e. applied before 30 June) in the 2019 application cycle.

If you have done better than expected, you can use the Adjustment system to look for universities that require higher grades. According to UCAS, 45 students used Adjustment to gain engineering places in the 2019 application cycle.

Figure 3 Timeline of applications

Entrance examinations

As competition for places is so fierce, some universities ask applicants to sit entrance examinations as part of the application process.

University of Cambridge

If you are applying for engineering courses at Cambridge you will be asked to sit the Engineering Admissions Assessment (ENGAA) in the November of the year you are applying. You will normally sit the test at your school or college and they will need to register you by 18.00 (UK time) on 15 October. The test has two sections, both multiple-choice questions on mathematics and physics: Section 1 lasts 80 minutes, has approximately 55 questions and allows a calculator; Section 2 lasts 40 minutes, has approximately 20 questions and no calculator is allowed. Information on the test can be found on the University of Cambridge website: www.undergraduate.study.cam.ac.uk/applying/admissions-assessments/pre-interview. You may also be required to sit the STEP (Sixth Term Examination Paper) Mathematics examination dependent on the college to which you are applying.

Sample University of Cambridge ENGAA question

Section 1
A cube has sides of unit length. What is the length of a line joining a vertex to the midpoint of one of the opposite faces (the dashed line in the diagram below)?

A $\sqrt{\frac{3}{2}}$

B $\sqrt{2}$

C $\sqrt{\frac{5}{2}}$

D $\sqrt{3}$

E $\sqrt{5}$

Answer: A

Taken from www.undergraduate.study.cam.ac.uk/files/publications/
engaa_s1_specimen_question_paper.pdf

University of Oxford

Engineering applicants for the University of Oxford have to sit the Physics Aptitude Test (PAT). The test is sat at your school or college in the first week of November. Your school or college will need to register you for the test by 18.00 (UK time) on 15 October.

The paper is two hours long and is divided into two sections: Section A, Mathematics for Physics (calculators allowed); and Section B, Physics.

Section B comprises short written answers. A typical question might involve being given the electric current recorded when a box containing electrical components with three terminals is connected in different ways to a power supply. You would have to deduce what the components are and the values of their electrical resistance from the data.

Sample tests and mark schemes can be found at: www2.physics.ox. ac.uk/study-here/undergraduates/applications/physics-aptitude-test-pat.

A good resource for sample and practice questions ahead of university interviews can be found at: http://i-want-to-study-engineering.org.

Other universities

It is usual for universities to require you to sit an admissions test for engineering, particularly where A level Mathematics is not present; however, certain universities may interview (although this may be subject to coronavirus restrictions). Interviews are usually not too formal and are more of a conversation surrounding your reasoning behind choosing an engineering course, as the admissions team want to ascertain if that is the right route for you. Kingston University, Liverpool John Moores University, Birmingham Metropolitan College and some other universities may ask some candidates to sit an aptitude or admissions test (in the course requirements information on the UCAS website, this is listed as 'IOT', which means 'institution's own test'). The tests are only normally required for mature students or students with unusual qualifications and you will be given information about whether you need to sit the test, and its format, when you apply. The tests are used as one piece of evidence in assessing candidates, alongside many other criteria such as grades achieved, predicted grades, the personal statement and references. The UCAS website provides a list of courses which may require additional admissions tests: www.ucas.com/how-it-all-works/explore-your-options/entry-requirements/admissions-tests.

Taking a gap year: deferred entry

Many students take a gap year between their final school or college examinations and the start of their university course. Universities are nearly always happy with this as students who take a year away from studies are often more motivated and mature when they start their degree studies. But in terms of the application, a gap year will only enhance your chances of getting a place if you use the year productively.

There are two application routes for students taking a gap year.

- You can apply for deferred entry; that is, you apply in the final year of the A level course for entry a year later. So, if you are sitting A levels in June 2021 you would apply for deferred entry in September/October 2022, not 2021.
- Alternatively, you can apply at the start of the gap year, once your A level results are known.

There are advantages to both routes, depending on your plans and A level (or equivalent) grades.

Advantages of deferred entry

- Once you have satisfied your offer, you will know where you will be studying in a year's time, and so you can make firm plans about what you will do during the year.
- You can plan to be overseas, either travelling or on work or voluntary placements, without worrying about having to return for university interviews.
- You have a second chance to apply to universities during the gap year if your initial application is unsuccessful.

Advantages of applying during the gap year

- More time to decide which field of engineering really interests you.
- You will already know your results so you can focus your application on those universities for which you have already achieved the necessary grades or scores.
- If your predicted grades are not high, but you feel confident of doing better than your school or college expects, you can avoid the possibility of being rejected on the basis of your predictions rather than actual ability.

'Applicants are led to believe that universities have differing views on the value of a gap year. This is not likely. A student who is taking a gap year to put off starting university and looks set to waste their time will always be frowned upon. But a student who has clear plans of almost any sort will gain much maturity from a gap year. Within a body of Freshers it is always possible to spot the students who have had a gap year. They are more mature and engage in interesting conversations beyond the obvious "What did you get in your A levels?"

'A typical gap year involves working for four to six months to earn money to fund travelling. This can be a good use of time. Some applicants engage in activities for charity which can be equally valuable. Anything that exposes you to new experiences will broaden your mind.

'A common concern with taking a gap year is that the student will forget their mathematics, which is such an important part of engineering, particularly in the first year. Even though it should be like riding a bike, there is no doubt that a year completely away from mathematics does mean that gap year students arrive a bit rusty and often do badly in the self-diagnostic tests that are commonly used at entry. But within a couple of weeks they are back up to speed with no long-term issues. So, this is no reason to avoid a gap year.

'Occasionally there are issues to be aware of associated with accommodation guarantees, etc. So, it is worth keeping in touch with the admissions tutor during the gap year. A significant minority of students complete their gap year and then decide to change degree course. Some universities probably use this as a reason for not encouraging gap years, but on the contrary this is a really good thing. If a student has had time to think about career options and has made a positive decision to change (see Chapter 7) then this is far better than arriving at a degree and realising halfway through the first year (or even later) that it may not be the right one.

'Keeping in touch with the admissions tutor can help both sides. You may learn some useful information that makes the transition to university easier. The admissions tutor is frequently one of your lecturers and will usually welcome the opportunity to get to know some of the students in advance of their arrival. Occasionally they can put you in touch with relevant student societies so that you are already in contact before the mad rush of Freshers' week. A good example is Engineers Without Borders that has branches at many universities and will welcome your attendance at some of their meetings before you arrive.

'Considerable value can be added by keeping a regular log of what you do during your gap year. When completing job applications five years later it can be very helpful to be able to reflect on what you learned from the year; trying to do it solely from memory leads to a less rich account.'

Lawrence Coates, Professor of Engineering,
University of East Anglia

If you are planning on taking a gap year, you must plan it carefully so that you will gain something from it. This could be life or work experience, maturity, a chance to extend your studies into new areas, independence or money to fund your university studies. A year spent resting and playing computer games, however attractive that may be after all your hard work at school, is not going to convince the universities that you will be a stronger candidate as a result.

Here is an excerpt from a personal statement about taking a gap year:

> *'I am taking a gap year in order to gain more maturity and experience.'*

Such a statement is not going to convince the admissions tutors that you have made constructive plans for your gap year, nor is it likely to help you develop or bring new skills and ideas onto their course. A better statement might be:

> *'During my gap year I am going to take part in a voluntary project building small-scale dams with Starfish, a voluntary organisation based in Thailand, in a mountainous area, to help control the water that causes soil erosion and crop damage. We will then work in a small village building toilet facilities for the old people who live there. I am looking forward to being practically involved with these engineering projects and to learn more about problem solving in real situations. To raise money for this trip, I will be working in my local hospital in the maintenance department.'*

This is much more impressive because the candidate has linked what she will do in her gap year to her future degree course (civil engineering), and it is clear that she has thought carefully about what she will do during the year.

A piece of advice: phrases such as 'I have arranged to ...' are much more convincing than 'I hope to ...' when discussing your gap year plans.

Student experience

'I initially wanted to go to university without taking a gap year, but during the weeks before my A level exams I decided against it and went for a deferred entry, as I hadn't got into a London university and wanted to give myself another opportunity. Taking a year out is significant, but there were a lot of things I wanted to do that I could not do while at university, mainly spending time in Colombia (where I was born) and starting a business.

'My first three months away were spent travelling within Colombia and spending time with family. After this period I settled in the capital city of Bogota, and established a routine which included a lot of meditation, fitness and salsa dancing. The remainder of the day I spent building and organising my business, which I had gotten a grasp of while travelling. I did travel a little bit, too, though sparingly.'

D.S. Barreto Romero, University College London, MEng
Mechanical Engineering with Business Finance

Gap year plans

It is always a good idea to check with your chosen universities that a gap year is acceptable to them before committing yourself. There is likely to be information on their websites addressing this. If there is none, you can email the admissions staff to ask them. This is particularly important if you are taking a gap year for reasons other than wanting to take a year between school studies and the degree course to gain experience in engineering.

Some of the other reasons for taking a gap year are:

- to work towards extra qualifications because you need to strengthen your application, or because you wish to change direction (for example, if you have studied mathematics up to GCSE only, you will be doing an evening class in A level Mathematics alongside your other projects)
- you started another course (such as a degree course in another subject; see Chapter 7) and then realised that it is not right for you, so you have withdrawn from it
- you have been working, and you now want to return to studying
- you may have had an illness or other issues that required you to take a break from studying.

Engineers like to solve problems. If there are no problems handily available, they will create their own problems.

Scott Adams, cartoonist

5 | Fire on all cylinders: The personal statement

Arguably, the most important part of your UCAS application is the personal statement. It is also the one part of the form where you have complete freedom to decide how you wish to demonstrate your suitability for the course to the selectors. You have 4,000 characters (47 lines) to convince your five chosen universities that:

- you have good reasons for studying engineering
- you have researched your future career thoroughly
- you have appropriate personal and academic qualities to become a successful engineer
- you will be able to contribute something to the department and to the university.

Who will read your personal statement?

Before you can write a personal statement you have to think carefully about your choice of courses. This is because each admissions tutor will read the personal statement with his or her own course in mind, and he or she will expect what you write to be consistent with the course. For this reason, make sure that the five courses you choose are closely linked in terms of course content and outcome. If you are applying for a civil engineering course, the selector will expect to read about your interest in civil engineering, books related to the subject that you have read, and relevant work experience. Similarly, an admissions tutor for electronic engineering will expect the personal statement to address this subject. Clearly, you cannot convince both that you are serious about their courses in one personal statement.

While there is a certain degree of crossover between some engineering disciplines, largely they are separate; therefore, it is better to be sure of your choice. If you were combining any engineering discipline in your application, the link would need to be strong in order to be able to reflect it in your personal statement, such as Mechanical Engineering and Automotive Engineering or Civil Engineering and Construction Engineering.

> 'Your personal statement is your chance to tell me what makes you special. I can already see your academic profile (past and predicted), so this is your opportunity to tell me how you're going to contribute to the Cardiff academic and social community. Is there something that makes you unique and you think we might value? Tell me about it and how you think this is relevant to studying engineering. Lots of students talk about how well rounded they are, and in the eyes of many admissions tutors, this is important. But once in a while I would like to read about an applicant who is truly special and unique. Have you done well in the face of adversity? Have you spent time in another culture? A leader? An athlete? How have these experiences made you better, and how will your skills help you to succeed and stand out amongst a group of other high-achieving peers? If you think it is relevant, tell me about it. Don't have anything like this to write about? That's OK too – maybe instead you can write about what challenges you expect to face in the transition to university and how you will overcome them.'
>
> Barry Sullivan, Head of Admissions, Cardiff School of Engineering, Cardiff University

Joint honours courses

Personal statements for joint honours courses are usually read by selectors from both of the departments to which you are applying. If you want to apply for a joint honours engineering and management course, someone from the engineering faculty will expect to read about engineering while their colleagues from the management department will want to read about management. This is fine if you apply to five similar courses, but if you apply for, say, three engineering with management and two single honours engineering courses you will find it difficult to satisfy the selectors from the two different types of course. Discuss your course choices with your careers adviser to assess how best to present these choices in your personal statement.

Applying for different courses at the same university

You can apply for more than one course at a particular university; you do not have to choose five different universities. But applying for two courses at the same university does not necessarily increase your chances of studying there. Take the case of a student who is desperate to study at a particular university, perhaps because she has friends there or she likes the city. She decides to apply for both the civil engineering and biomedical engineering courses. The admissions tutor for civil engineering will look at her application for the civil engineering course, and the admissions tutor for biomedical engineering will look at the application for this course.

Our applicant's main interest is civil engineering, so her personal statement emphasises this, but it also devotes one paragraph to her interest in biomedical issues. The civil engineering admissions tutor reading the personal statement will judge it on how it addresses this course, so he or she might not be fully convinced that the student is serious because the personal statement will not focus enough on the reasons for the choice of civil engineering, and what the candidate has done to investigate it (reading, work experience, etc.). Similarly, biomedical engineering is a very specialist area, and the admissions tutor for this course will expect to read a personal statement that focuses on this, and he or she is not going to be very interested in reading about bridges and roads.

So, by trying to give herself a better chance of getting to this university, the applicant is actually reducing her chances. There are some instances where it is possible to apply for two separate courses at the same university, if they are very similar, but it would be advisable to discuss this with the university admissions department before doing so. So, when writing the personal statement, try to imagine how it will come across to each of the departments to which you are applying. Do not try to write something too general in order to allow yourself the luxury of applying to a wider range of courses.

The structure of the personal statement

What is a perfect personal statement? Of course, there is no such thing. The key to writing a personal statement is to think about the word 'personal' – it is about you and so it has to reflect your strengths, achievements, qualities, research and ambitions. Having said that, there are some elements that are important in creating a successful personal statement:

- reasons for your choice of course
- what you have done to investigate the course and the profession
- what unique qualities and achievements you have
- other things that may affect your application, such as gap year plans
- make it pertinent, relevant, concise and personal.

Reasons for your choice of course

These could include:

- what first got you thinking about engineering, for example watching the news about a new engineering project, an article in a newspaper about new developments in the mobile phone industry, or personal experience such as work experience – this could date back many years, for example how you liked taking things apart and reassembling them when you were very young
- things you have studied at school, for example a topic in physics or chemistry that particularly interested you
- how your particular interests and qualities make engineering a suitable career.

What you have done to investigate engineering

This could include:

- books, articles, magazines or websites that you have read (anything here must be contextualised and the emphasis be on what you have learnt)
- work experience (see Chapter 2)
- talks or lectures
- school visits to museums or exhibitions
- the relevance to engineering of things you have studied at school.

Your achievements

These could include:

- academic achievements; for example, Maths Challenge or the Mathematical Olympiad
- extracurricular activities and accomplishments
- responsibilities; for example, sports captain, head of house, chairman of a club or society
- hobbies.

Your personal qualities

These could include:

- teamwork; for example, sports, school orchestra, Management Challenge, Duke of Edinburgh expeditions
- communication skills; for example, voluntary work, school prefect, school council
- leadership; for example, being captain of the school rugby team
- problem solving; for example, competitions, experimental projects, EPQs.

Other information relevant to the application

This could include:

- gap year plans
- if you are an international student, why you want to study in the UK
- what qualities or experiences you could offer to the university or the course.

Blogs and online lectures

At the time of writing, the world is in the midst of a global pandemic. As such, virtual open days and tours are now commonplace. Technology has come to the rescue with strong, interactive platforms available to roughly replicate the normal experience of university teaching. Many universities upload lectures so that they are available to the public (including potential students). If you are unable to go to talks and presentations because of school commitments or because you do not live near a university that offers these, then downloading online lectures or following blogs is a good way to learn more about engineering.

Two good examples (Imperial College London and the University of Bath) are given below (www.imperial.ac.uk/be-inspired/social-and-multimedia/lectures-online   and   https://soundcloud.com/uniofbath/sets/public-lecture).

'Preparation for writing the personal statement

1. **Evidence of study of what the subject involves.** This can be anything from indicating some relevant books or websites to having gained work experience with a local company. Good websites, such as the Royal Academy of Engineering, have a lot of useful and reliable explanations about what engineering involves. Increasingly, applicants have been attracted to engineering by watching various programmes on the Discovery Channel, such as 'Extreme Engineering'; this is fine.

2. **Courses and summer schools.** A student who has made the effort to attend a summer school or Headstart course has demonstrated that they have tried to find out what a degree involves. This isn't necessarily the same as what a career involves, but is a good start.

3. **Evidence of an understanding of the breadth of the subject.** Most engineering disciplines are associated in the popular imagination with specific elements, for example mechanical engineering with Formula 1, or civil engineering with structural

---

design. Any evidence of understanding that mechanical engineers also study thermodynamics, efficient engines, sustainable design, etc., or that civil engineers are particularly concerned with water treatment, flow and management, soil mechanics and geotechnics and are involved in people management, will demonstrate some depth to the background research.

4. **Passion.** It might be argued that an applicant doesn't know enough about the subject to yet be passionate about it. This might have been true 20 years ago, but the internet and YouTube have put an end to that. Any applicant interested in engineering can find a wealth of informative resources that could excite their interest.

5. **Understanding the value of transferable skills.** Sometimes applicants approach the exercise in too narrow a fashion. Usually this manifests itself as a focus on the technical side of engineering and demonstrates a lack of awareness of the broad transferable skills that are needed and developed in a degree. Accordingly, almost any work experience can offer valuable insights into the applicant's approach to team-working and problem-solving. It will not be enough to mention in some vague way that you like working in a team; it would be far better to explain how a specific team-based activity has helped you to understand your strengths and weaknesses.

6. **The value of work experience.** Often applicants are reluctant to mention work experience unless they think it is directly relevant. I have known applicants not mention a Saturday job because it is in a supermarket. However, any job demonstrates experience of the world of work and offers opportunities for the curious student to find out how the business works, even if they are just shelf-stacking.

7. **Studying A levels or equivalent requires commitment.** So, any applicant who is managing to do well at this level and is still maintaining committed contact with organisations such as Scouts, local sports teams or musical productions, for example, is demonstrating the necessary ability to work hard and play hard that is required at university.

8. **Think very carefully about what to write about leisure activities.** You have to remain honest, but if all you do is play games and socialise, now might be a good time to start broadening your activities. Some applicants make the effort to go to local branch meetings of professional engineering societies,

others subscribe to industry magazines. All such things demonstrate some commitment to the career. Remember, you are only at university for three or four years; your engineering career is for life. You need to expose yourself to the industry in some way to find out what the job involves.'

Lawrence Coates, Professor of Engineering,
University of East Anglia

Advice from a student

'When writing a personal statement for your UCAS application, my advice would be to first write everything you can think of that is remotely relevant, including every reason you have for wanting to study your chosen subject and all the extracurricular activities you've done. From there, you can begin to redraft and remove the things you think are less useful to your application. Bear in mind that the majority of your personal statement should be focused on academic issues and only a paragraph should be given to personal hobbies, such as sport and music. When talking about non-academic activities, you should try to explain the skills that this activity has helped you develop. For example, if you play a musical instrument you can explain that it has helped you gain personal discipline – you should explain to the reader how your activities have made you into a better university candidate. Ensure your personal statement is true and that you've read everything you said you had. You should have notes on the books you've read which you can use to refresh yourself on their contents before your interview so that when you are asked about it you have something intelligent to say.

'When preparing for an interview you should always do a mock interview with someone you don't know well to best simulate the environment in which you will find yourself. Also, to prepare for an interview, you should write out model answers to some of the stock interview questions, such as "Why do you want to study this subject?", "Why do you want to study at this university?" and "Why are you suited to study this subject?" Remember, you should not rush your UCAS application, so start preparing as early as you can to maximise your chances.'

Jordan Massiah, Magdalene College, University of Cambridge,
MEng Engineering

Analysing a personal statement

I first became interested in engineering during my A level Physics classes [1]. As a child I had always enjoyed taking things apart, although I was not always able to put them back together again.

To further investigate engineering, I spent a week at a local architecture practice, looking at how they worked with a structural engineer to ensure that their designs were practical. I also visited a construction project as part of my school's work experience programme, and I was fascinated to see how things I had read about and studied were used in real life [2]. I have also read as much as I can about engineering, including *A Short History of Engineering Materials* by John Cameron [3]. I have enjoyed going to public lectures on engineering, and talking to engineers, who have given me a much better idea about a career as an engineer [4].

Alongside physics, I am studying mathematics and history of art. Mathematics is an important tool for engineers [5], as well as teaching me to think in a logical way. History of art is an analytical subject, and it puts architecture within a social and political context. It also involves looking at the use of materials, and how architecture and sculpture were able to develop as new materials were introduced [6].

I enjoy sport and music. I am captain of my school football team and so have had to develop leadership and communication skills as well as physical fitness. I play the guitar in a band, and the cello in the school orchestra, which help me with my manual dexterity and teamwork. Outside of school, I enjoy cooking and cycling. I am a member of my local cycling club and compete most weekends.

I believe that my combination of A levels and my research into engineering as a career make engineering an ideal choice for me.

Points raised by this personal statement

An admissions tutor who read this sample personal statement made the following points.

General

While it is clear that the candidate has done some research, there is very little detail in the statement – it is very general – and so I do not really get a clear picture of the depth of knowledge the candidate has about engineering, or about his/her particular areas of interest. It is also, to be honest, a little on the bland side and also a bit frustrating – a lot of sentences which should lead to something that will interest me end without giving me any information.

Specific points (the numbers refer to the relevant passages in the statement)

1. It would have been nice if he/she could have given an example – perhaps it was electricity, or the behaviour of materials, or some problems involving forces?
2. This should have been the most interesting part of the statement. The candidate could have told me about the links between his/her studies and the work experience. This will tell me that the student has gained something from the work experience, and that he/she is thinking about engineering rather than just doing work experience to look good on the application form.
3. I am always encouraged when students read around the subject, but what I would like to know is, again, some detail. How do the ideas in the book link to A level study and the real world?
4. This would have been an ideal opportunity for the candidate to show me that he/she has been really thinking hard about his/her future career and about whether he/she has the right qualities to be a successful engineer.
5. Give an example.
6. Not many applicants for engineering study history of art, so this immediately makes him/her stand out. And I don't know much about history of art, so an example here would be interesting for me to learn about – I'm sure it would make me want to learn more by meeting the candidate.

A revised (and much better) version of the personal statement, based on the above advice, is now given.

Revised sample personal statement

As a child I had always enjoyed taking things apart, although I was not always able to put them back together again. I first became seriously interested in engineering during my A level Physics classes when we looked at the properties of solid materials, and I began to understand why, for example, the development of reinforced concrete revolutionised the construction industry.

To further investigate engineering, I spent a week at a local architecture practice, looking at how they worked with a structural engineer to ensure that their designs were practical. I also visited a construction project as part of my school's work experience programme, and I was fascinated to see how things I had read

---

about and studied were used in real life. The engineers explained that the concrete girders used to hang the curtain walls of the office block had to be strengthened along their top surfaces because in a cantilever, the tensile forces are at the top and concrete is weaker in tension than it is in compression. I have also read as much as I can about engineering, including *A Short History of Engineering Materials* by John Cameron, and I was fascinated about how the use of cast iron in early 20th-century America enabled architects and engineers to build the prototypes to today's skyscrapers. I have enjoyed going to public lectures on engineering, and talking to engineers, who have given me a much better idea about a career as an engineer. In particular, I began to understand that an engineer needs to be able to analyse information quickly and to be able to solve problems. My aim is to study structural engineering and then to work alongside archi-tects in the creation of exciting new buildings.

Alongside physics, I am studying mathematics and history of art. Mathematics is an important tool for engineers because the start-ing point of any engineering project is an analysis of its feasibility. The use of integration to find a centre of mass, for example, can help with the design of an asymmetric building. History of art is an analytical subject, and puts architecture within a social and politi-cal context. It also involves looking at the use of materials, and how architecture and sculpture were able to develop as new materials were introduced. The transition from the small, squat Romanesque churches to tall and graceful gothic cathedrals in Europe was due to the invention of the flying buttresses, which, in turn, were only effective when tensile forces were reduced by the addition of heavy statues or decorative stone elements.

I enjoy sport and music. I am captain of my school football team and so have had to develop leadership and communication skills as well as physical fitness. I play the guitar in a band, and the cello in the school orchestra, which help me with my manual dexterity and teamwork. Outside of school, I enjoy cooking and cycling. I am a member of my local cycling club and compete most weekends.

I believe that my combination of A levels and my research into engineering as a career make engineering an ideal choice for me.

Adding the extra information requested by this admissions tutor would add detail, make it more interesting for him to read (so he is more likely to want to meet the student), demonstrate that the student is interested enough in the subject to be thinking about links between his studies and what he has experienced, and bring it up to the required length.

Linking your interests and experiences

In your personal statement, try to avoid creating what amounts to a list of things you have read, studied or experienced. It is better to make connections between these in order to demonstrate that you have thought carefully about what is required to be a successful engineer:

- how your A level studies are related to things you will study at degree level
- how something you observed during work experience stimulated further reading
- how skills that you gained from extracurricular activities, such as communication or leadership, are useful for potential engineers.

You could link:

- an article you read about increasing the battery life of mobile phones with something you studied about chemical reactions in A level Chemistry
- the design of a new building with your study of forces in physics
- the need for engineers to be good communicators with your role as your class representative at school
- the follow-up research you did on wind turbines with an on-site visit on a school trip
- a news story on a new aircraft design with an article in the *New Scientist* about composite materials.

Work experience

Work experience is important as it demonstrates a commitment to the subject outside the classroom. Remember to include any experience, paid or voluntary. If you have had relevant work experience, mention it on your form. Explain concisely what your job entailed and what you got out of the whole experience. Even if you have not been able to get work experience, if you have spoken to anyone in engineering about their job it is worth mentioning as all this information builds up a picture of someone who is keen and has done some research. See Chapter 2 for further information.

How to get started on the personal statement

Different people approach this in various ways. If you are struggling to get started, begin at the end or in the middle. To start at the end is often a good way to write this statement as the summation should be your sales pitch as to why you want to take this course. If you know that, you can work your way through the rest of the statement. If you treat these as isolated paragraphs then you will be able to access them more easily. Another good strategy is to start by making lists of anything that you think is relevant to your application. Then begin to organise them into sections. Your personal statement could include some of the following points.

I first became interested in engineering because …

- I read an article in a newspaper about …
- I read the book '…'
- I saw a piece on the news about …
- of my work experience
- of my father's job
- of something I have enjoyed studying at school.

I have investigated engineering by …

- reading books
- reading the *New Scientist*
- reading the Royal Academy of Engineering website
- work experience
- going to a public lecture at a university
- discussing engineering with an engineer
- downloading a podcast of a university lecture on iTunes U.

From my work experience I learned …

- that the qualities necessary to become a successful engineer are …
- how the theory we study at A level is applied practically
- the importance of communication skills/problem-solving/leadership …

Other points to include:

- a particular A level topic is useful because …
- my part-time job is useful because …
- my role as rugby captain has taught me …
- being leader of the school orchestra has taught me … (or lead in the school play, or …)
- during my gap year I will be …
- I was awarded first prize for …

Only when you have the ideas listed should you start to write full sentences and to link the points.

Many people find that the best way to make it fit the available space is to start with a statement that is perhaps double the allowed word count but that incorporates everything you think is relevant, and then start to edit it down by:

1. removing any superfluous adjectives or complicated sentence structures, e.g. 'I was lucky enough to attend a stimulating talk on the exciting topic of wind turbines delivered by the renowned engineer Richard Martin, which inspired me to read xxx', could more succinctly be written 'I read xxx after hearing a talk on wind turbines by Richard Martin'

2. removing any passages that are not about you. Often, I read personal statements that try to impress the admissions tutors by explaining in detail engineering concepts that they have read about. The chances are that the person reading the statement is far more familiar with the topic than you are so will skip this section completely and so you have wasted valuable space which could have been used to tell him/her more about you

3. removing sentences that state the obvious: 'I chose physics and mathematics at A level because these will be useful when I study engineering'

4. removing passages that contain nothing factual, particularly closing statements such as 'I believe that my commitment to engineering coupled with my academic achievements and ability to work hard will make me an asset to your esteemed institution.'

Once you have gone through this process you should be getting near a personal statement that is around the right length. Then you can fine-tune it so that it fits exactly. A useful tip is to ask your referee to include anything that you have had to discard in his/her reference. So, if you have run out of space and have to omit information about, say, your role in organising a school event, then the referee can incorporate this into the Reference section of the form.

Things to avoid at all costs

- Boasting about your achievements.
- Telling the admissions tutor why he/she should offer you a place.
- Quotes from famous engineers, scientists or philosophers.
- Listing your examination results or predicted grades (these appear elsewhere on the UCAS form).
- Adapting or copying personal statements that you have found on the internet or in books (including this one!).
- Spelling mistakes or grammatical errors.

First impressions

Engineers tend to be practical people who can convey complicated information in ways that are easy to understand. They are problem-solvers who can get to the heart of an issue, discarding information that is not relevant to the issue they are dealing with. They understand the importance of turning ideas into solutions that have a benefit to others. Bear this in mind when writing the personal statement. Make the statement easy to read by using simple sentences, adding spaces between paragraphs, and sticking to the point. Give it a coherent structure and a logical flow. Remove anything that does not have a practical use. Treat the statement as you would an engineering project. Its purpose is to get you the offer at your chosen university.

Language

It is important that you use succinct language in your personal statement and make every word count. Remember that you are limited on the number of characters you may use so it is important not to use up this vital space with superfluous language. Keep it as simple and clear as you can, rather than using overcomplicated language in an effort to impress.

For example:

> *'I was privileged to be able to undertake an internship with a well-known engineering company where I was able to see the benefit of having the ability to be confident with information technology'* – approximately 200 characters.

Could be rewritten as:

> *'My three weeks' work placement at Rolls-Royce showed me the importance of being proficient in using spreadsheets'* – approximately 110 characters.

Similarly:

> *'I was honoured to be chosen to play the lead role in my most recent school drama production, and interacting with the producer and the rest of the cast involved a significant degree of communication and teamwork'* – approximately 210 characters.

Could be rewritten as:

> *'Performing as the lead in my school play taught me to work and communicate effectively with others'* – approximately 100 characters.

Phrases to avoid

- 'It was an honour to …'
- 'I was privileged to …'
- 'From an early age …'
- 'For as long as I can remember, I have dreamt of …'

'Writing the personal statement

- **Accuracy of spelling and grammar.** The current jobs market is such that companies are using software that automatically rejects applicants after three spelling mistakes. So it is important to demonstrate this attention to detail in a personal statement. The statement should not have any mistakes at all when it is finally polished and submitted. This probably means it will have been through 20 draft versions. It conveys a lot to an admissions tutor when an applicant has made a spelling or grammatical error, or has doubled or missed out words. If engineers don't pay attention to detail, there could be serious repercussions in most engineering disciplines, including safety issues.
- **Written communication.** Most practising engineers spend a lot of their daily working life writing good English in technical reports. Most applicants study mathematics, physics and chemistry and hence do not have formal opportunities to improve their written English. So, any evidence that an applicant has attempted to broaden their reading beyond textbooks or involved themselves in extended writing will be useful.
- **Structure.** Some personal statements are chaotic and rambling. Good ones have obviously been planned and tell a story about the applicant in a logical order. It is important for an applicant to realise how much time it takes to prepare a statement.
- **Peer review.** An essential step in constructing a personal statement is to let a close friend look at it. Frequently they will notice some important aspect of your life that you have forgotten to mention because it is so obvious.
- **The personal statement can steer the interview.** Some universities use the personal statement as a basis for allocating applicants to staff for interviews. So it is important to only include things that you want to discuss in these circumstances. So, including things because you think it will look good, but not knowing much about them, is a bad idea.
- **The personal statement must be personal.** It is usually quite obvious if an applicant is trying to guess what the reader wants. It is far better to write from the heart, demonstrating passion for the subject. So many personal statements look the same. So, when an admissions tutor reads one with some interesting element that nobody else has mentioned it livens it up. Start by thinking of five things that could only be included in *your* personal statement; in other words, make sure it is personal not formulaic.'

Lawrence Coates, Professor of Engineering,
University of East Anglia

Sample personal statements

The following examples of personal statements have been contributed by students who were successful in gaining offers from their chosen universities.

Please remember that these are personal statements; that is, they reflect the experiences and ambitions of the students who wrote them – so **do not attempt to copy them or adapt them for your own use**.

Personal statement 1

Engineering is one of the foundations on which modern civilisation is built, and the world we live in today has emerged as a direct result of this application of science to address problems and develop innovative and often simple solutions to seemingly complex issues.

I am a regular reader of IEEE *Spectrum* and *Professional Engineering*. These have given me insight into current innovations and allowed me to see a hint of the many boundaries being pushed by engineering today. The book *To Engineer is Human* has inspired me in my pursuit to be an engineer and has introduced me to what it means to be an engineer and to understand some of the pivotal success and failures in engineering history.

The concept of futureproofing appears to be long defunct, and the exponential rate of technology development excites me. From my interest in hardware engineering, I built and modded my own PCs, gaining awareness of the rapid advances in technology and the challenges we face in keeping up with this relentless pace. These are the challenges that motivate me to learn and understand more.

I also took advantage of access to CAD software and a fused filament deposition 3D printer to design and develop a mini quadcopter controlled by an Arduino microcontroller. This required research into materials, programming of the microcontroller, frame and rotor design, and parts assembly. After visiting Philips Avent, a large manufacturing company in Suffolk, I was left in awe by the advanced manufacturing techniques used for rapid prototyping. Seeing a machine performing additive manufacture by laser sintering completely engaged me and made me want to be part of this ongoing engineering and manufacturing revolution.

I enjoy solving theoretical problems, and find myself asking more to satisfy my hunger for knowledge. Chemistry allows me to appreciate the complexity of modern-day engineering materials

---

and understand the benefits of high-performance alloys and ceramics; it fascinates me that viewing intricate problems at microscopic level can yield solutions on a macroscopic scale. My study of mathematics has improved my ability to analytically and methodically solve problems; my participation in several UKMT maths challenges is testament to this. These challenges helped me consider different and novel approaches to problem solving other than those I had used before. Studying physics has fuelled my interest in the intricate connections within our world, such as the details essential for safe nuclear fission and further development in the successful use of nuclear fusion. This led to my decision to read *Six Easy Pieces* by Richard Feynma, which pays attention to numerous intricacies and caveats within projects and everyday life.

I was fortunate enough to take part in a work experience scheme in the R&D lab of a chemical firm. This highlighted how mechanical systems are intertwined into all types of industries. I also learnt the importance of efficiency and health and safety. Although intricate fail-safes incorporated into machinery design are fascinating, something as basic as a kill-switch on a high temperature furnace door made me think even more about considering the more complex mechanisms and trying to understand them by breaking them down into their basic components.

Captaining a rugby team and being an active member of my local hockey and athletic clubs have allowed me to develop both my leadership, communication and organisation skills while maintaining a healthy lifestyle. I believe these attributes are key for even the smallest scale engineering projects. I have also demonstrated the importance and use of these skill sets during my participation in both Bronze and Gold Duke of Edinburgh awards.

I look forward to having the opportunity to delve further into the multi-faceted field of mechanical engineering and join the next generation of engineers steering the direction of our future.

Personal statement 2

Working for Thai Toshiba Electric Industries exposed me to the extensive nature of electricity and electronics, as well as engineering as a career. It reaffirmed my decision to apply for electrical and electronic engineering (EEE). As a child, I loved Tamiya remote-controlled electrical cars, and since my first experience of constructing a PCB seven years ago, I have become increasingly captivated by electricity and electronics. Reading about recent developments of efficiency in smartphones and electrical cars has made me realise the massive potential of technological break-throughs in this area.

From attending the Headstart Electronics Course at the University of Bristol, I gained valuable experience through a series of mini projects and lectures. The course developed my knowledge of electronics and engineering, boosting my passion for the subject. A work-shadowing placement with the factory manager at Bangkok Cable Company increased my understanding of electrical engineering on an industrial scale and the intellectual approaches of an engineer. This led to my achievement of winning the Gold CREST Award for the QinetiQ Young Engineers Project by producing a navigation aid for the visually impaired. This com-petition tested my creativity, logical thinking and motivation.

Last summer, work experience at Type One Exhaust Performance, a custom car exhaust company in Thailand, gave me the opportu-nity to further my knowledge of car systems, where electronics and technology play a substantial role in modern vehicles. I was introduced to the design process of each custom exhaust system and the engineering aspects of these designs to maximise efficiency,

Studying mathematics and further mathematics has enhanced my ability to solve problems logically, equipping me with the versatility through applied modules to tackle the mathematical elements of the course. Reading James L. Adams' *Flying Buttresses, Entropy, and O-Rings: The World of an Engineer* gave me a taste of the engineering discipline and furthered my appreciation of the crucial role mathematics and science plays in engineering.

Another parallel element of engineering is physics, where I am most drawn towards the electricity-related spheres, such as elec-tromagnetic induction and superconductors due to their fascinating nature and role in technology. David J.C. MacKay's *Sustainable Energy – Without the Hot Air* emphasised the current energy cri-sis and the importance of the contribution electrical engineering plays in solving a global issue; reading this has inspired me to take up a full-time work placement at Bangkok Solar Company after my exams, where I hope to extend my knowledge of solar panels as a renewable resource.

My academic success, reflected by my Honorary Academic Scholarship, shows that I have the capabilities and determination required to study EEE. Beyond the classroom, having been appointed managing director of my Young Enterprise group, as well as lead researcher and deputy lead designer for my Gold CREST award team, I have extended my skills in management and group discussions. Furthermore, as a peer mentor, I have learned how to interact with different people, which is a fundamental part of everyday life. Being part of my school's senior rugby and tennis teams developed my abilities to compete at a high level and perform under pressure. These various activities emphasise my good time management and teamwork skills alongside my commitment.

EEE appeals strongly to my passion for technology. It is a dynamic discipline that utilises my enjoyment and strengths in mathematics and physics, applying these skills to realistic problems and the idiosyncrasies of working as an engineer. The diverse scope and possibilities of this field and designing projects to fit the framework of the modern world fascinates me.

My ultimate ambition is to pursue a career at the forefront of the electrical energy and electronics industry and to improve technology for developing countries more efficiently and sustainably.

Personal statement 3

Meccano is a great way to start leaning about engineering. So much so that by the time I was 10 years old, I raced and maintained my own Rotax go-kart. At the age of 11, I had winched up our 22-year-old jet ski, changed the supercharger clutch, then added an intercooler with a fan and fuel-mixture adjuster; the purpose, to increase power. Since then, I have modified a Jeep JK Wrangler (with a lift kit, fuel system work and upgrade, a supercharger and lots of suspension components, as well as customised body work), rebuilt a Dodge Ram 1500 pickup, and a Mercruiser 5.7L V8 speedboat engine. I have also built a modified 1986 BMW 3 Series, using a 2.5L twin turbo Toyota 1JZ-GTE engine and adapting the engine by changing the internals, turbos and engine management. I am very keen as a result to study Automotive Vehicle Design and Performance and I particularly look forward to using CAD tools such as CATIA and Ansys in this, which I have only read about so far.

I love to read about the latest scientific and technological innovations and have been especially interested of late in the idea of space tourism after Sir Richard Branson's failed attempts to push this frontier. It is extraordinary to think that such an idea is within our lifetime and it would be a remarkable feat of engineering and imagination. What the space race underlines for me though is that engineering is a team-based career across the profession, and while there is competition between firms, it takes one firm to trailblaze an idea for another firm to develop it further, all for the better of society. It is a sign of the advancements in this field that the land speed record is expected to be broken at every attempt. In 1904, when William Vanderbilt broke the land speed record at 92mph, this was seen as an incredible feat, which quantifiably at the time it was. Today, Bloodhound SSC are planning to achieve speeds of 1,000mph in a car designed around the engine of a Eurofighter Typhoon. Not only is that a sign of the times but an indication of man's ability to adapt, to develop, to innovate. We are now at the stage whereby it is realistic to say anything is possible and I want to be at the forefront of such a profession.

After my A levels, I was accepted onto the Engineering Foundation programme at City University. This has allowed me to build on my A level knowledge and apply my physics in real world examples. Studying heat transfer introduced me to the idea of thermofluids and entropy, which is something I look forward to learning about in more detail at undergraduate level. Physics has taught me about materials used in design, while mathematics will help with the programming elements of modern engineering.

In order to develop my understanding of this field, I have undertaken internships which have helped me learn how to adapt my self-taught interest to a future career. In 2013, I worked at a Mercedes Service Centre and Repair Shop in Cairo, where I repaired and serviced cars. I learned new procedures; how to top-up AC systems with Freon, how to adjust ignition timing and how to take apart, reassemble and change parts in both manual and automatic transmissions.

In 2014, I worked at Aston Martin's newest state-of-the-art service centre in Dubai. Here I memorised the different models, variations between them, and the details of spare parts. I learned wheel alignment, how to apply diagnostics and to balance wheels with carbon ceramic brakes, as well as how to buff a car's paint-job. Most crucially I learnt the team-based aspects of this job, understanding that it is the sum of all parts that makes something work, and was commended for the role I played.

Throughout my life, I have learnt to be organised, disciplined, timely and responsible. I have shown in my studies that I can approach a problem in a systematic way and troubleshoot with good results, all of which are the marks of a successful engineer.

Personal statement 4

I have been keen to study engineering from a young age. This is because I have always been interested in how things work, from building Lego structures when I was younger to disassembling my laptop more recently. I believe that engineering is a vital part of society today, and I am eager to specialise in the subject with a specific interest in aeronautics. I would find an engineering degree a stimulating experience which would allow me to further learn how things work.

At school I have enjoyed learning sciences, and am studying mathematics, physics and chemistry at A Level. I have always liked mathematics and when I was 14 I achieved 150th place in the Cayley Paper. I have a preference for applied mathematics, such as mechanics, as I feel as though the work is more relevant to real-life situations. From studying physics, I have realised that things I have taken for granted are actually more complex. For example, GPS systems in cars wouldn't work if the engineers who built the technology didn't take into account Albert Einstein's Theory of Relativity. The effect of relativity is that time runs slightly faster on a satellite than on the Earth, leading to measurement errors of about ten kilometres. It fascinates me that something as seemingly mundane as a satnav is affected by relativity, and I look forward to being able to use my knowledge of physics to help design my own creations.

In chemistry I found the topic of CFCs particularly interesting, as it gave me insight into why these chemicals are so harmful to the ozone layer. I also liked learning about green chemistry and what chemists are doing to try to control emissions of polluting gases. At AS level I studied economics and have learnt how to use various analytical techniques to interpret data. I performed strongly in my GCSEs, achieving a high standard in all subjects.

I follow recent developments in the scientific community and read scientific journals such as *New Scientist*. Recently I read Stephen Hawking's *The Grand Design*, which I found a challenging but entertaining read on the frontiers of modern physics. I found Hawking's description of Conway's Game of Life particularly interesting. The laws of the game are very simple, but complex behaviour can arise, e.g. gliders. As a result, to someone living in the game the laws might appear complicated. I found this different perspective to a set of laws exciting as it is comparable with the laws of physics, where we are inside the 'game' so the laws appear complicated. I am now viewing technological designs with new perspectives to further understand how they work.

Aside from my academic interests, I enjoy riding my racing bike. Forcing myself up a hill while ignoring the pain in my legs focuses my mind, as all I think about is the next push on the pedals. I find cycling to be a great way to exercise and as a way to challenge myself mentally.

Over the last year, I have been volunteering for Lyndon Arts Trust, helping out in an after-school art club at a primary school. I support the children's ideas by giving suggestions for their projects, while developing my own project at the same time. I have learnt valuable lessons of how to work in a team and how to work in a supporting role. I have also had the responsibility of being a prefect which has improved my time management and leadership skills.

Engineering is a subject that has captivated my mind. I am motivated and excited by the prospect of becoming an engineer and relish the next stage of my education and life at university.

Personal statement 5

Family trips to Formula 1 races gave me an early appreciation of how maximising engine output, optimising aerodynamics and using electronic systems impact racing performance. Such real-life science applications are particularly appealing to my inquisitive and problem-solving nature and sparked my interest in engineering. It was inspiring to grasp, through J.B. Williams' insightful history of electronics advances in *The Electronics Revolution* how seemingly unlinked inventions used fundamental science and each other to collectively revolutionise our world.

Academically, I channelled my excitement for engineering into A-level Mathematics, Physics and Chemistry. I am captivated by how mathematical concepts, such as differential equations, form the basis of scientific endeavour. I am also fascinated by the idea of covert mathematics in everyday life, as Jordan Ellenberg engagingly describes in *How Not to Be Wrong*, such as by employing correlation and causality to examine the relationship between lung cancer and smoking. Physics and chemistry provide me with valuable practical and analytical skills, and teach me how advanced scientific knowledge can be innovatively utilised in life-enhancing products and services. For example, on a trip to CERN, I learned how particle physics research was applied to semi-conductors, the world wide web, medical scans and sterilisation.

My school activities are equally science-centric, and I regularly attend STEMsoc events. I am a member of the Robotics Club and our team won a bronze medal in a London Schools competition, where we had to build and program a robot to undertake a variety of timed tasks. In DVA Club, I improve my practical skills, and I designed and constructed a laser-cut Perspex clock and a wood-inlay chess-board table. I made a detailed poster of an experiment on static pressure changes in a sloped water streamline and presented a study on the exciting properties and wide applications of Aerogels to an audience of teachers and students. In a sixth form extended essay, I discussed how Formula 1 innovations benefit road cars.

I enjoy reading MIT's *Technology Review* and attending Royal Society and IET lectures, such as 'The New Ford Focus' on how multi-disciplinary engineering teams designed new features, improving the car's functionality and safety. Four summer placements solidified my motivation to study engineering. At Vauxhall's Luton factory, I worked in the steel presses, body shop, general assembly, paint shop and quality control station, and presented a process flowchart for Vivaro vans to factory managers. At BMW Mini's Cowley factory, I monitored spray-painting robots and repaired assembly-line disruptions, carrying out concurrent tasks with different teams. At Arup, I learned independently to use Rhino software to model a football stadium's airflow and took the initiative to study Python. At Troika, I explored how 4D digital content is computer generated. My time at BMW and Vauxhall was particularly useful for getting a holistic perspective of how engineering plants operate.

Beyond academics, I aim to be a rounded person and eagerly mentor younger students. During Combined Cadet Force, I exhibited discipline, organisation and self-management skills, earning my school's 'Best Cadet' award. My community involvement includes organising charity events at our church and participating in my school's 'rowathons'. I also volunteer at Oxfam's Southern Logistics Centre to sort donations, perform on-line shop functions and fulfil sales. I am a keen sportsman, representing my school in football and squash. I play the guitar and have been an enthusiastic member of the choirs of my school and church since 2011.

I have enjoyed cultivating my childhood curiosity to build a strong sciences foundation for undergraduate studies. I relish the challenge of attending a demanding and stimulating course and eagerly anticipate fulfilling my ambition to forge a creative career in electrical engineering.

General tips for completing your UCAS application

- Before submitting it, also ensure you check your application through very carefully for careless errors that are harder to see on screen.
- Keep a copy of your UCAS application so you can remind yourself what you wrote, prior to an interview.
- Ensure that you have actually done all the things you mentioned in the statement by the time you are interviewed.
- Research the full course content, not just the first year.
- Research the entry requirements.
- Ask your teachers for your grade predictions.
- Ensure your personal statement is directed at the courses you are applying for.
- Include lots of detail in the personal statement.
- Get someone else to proofread your personal statement.
- Illustrate your points with examples and evidence.
- Do not waste valuable space in the personal statement – make every word count.

Joke

Two antennas got married – the wedding was lousy, but the reception was outstanding.

6 | A well-oiled machine: Succeeding at interview

Not all universities interview candidates, but it is likely that at least one of your five choices will do so, so you need to be prepared. Your interview will decide whether you will be offered a place or not. The information on the UCAS application will have been the basis on which the decision to interview you was made, but a good UCAS form cannot help after a poor interview. So prepare thoroughly. Here are some important points to consider.

- If you interview well, and you subsequently narrowly miss the grades that you need to take the place, you may still be offered the place.
- Interviews are normally conducted in an informal and relaxed manner, the purpose of which is to allow you to talk about your interests and suitability for the course.
- Think about the impression you will make – think about your body language, eye contact and communication skills.
- Go into the interview with a mental checklist of what points you wish to mention and try to steer the interview to address these (see below).
- Interviewers are less interested in investigating your subject knowledge than in looking at how suitable and committed you are for their course. So, evidence of research and appropriate qualities such as analytical or problem-solving skills are important elements of a successful interview.
- Remember, your future teacher or lecturer might be among the people interviewing you. Enthusiasm and a genuine commitment to your subject are extremely important attitudes to convey.
- An ability to think on your feet is vital – engineering is about problem solving. Don't try to memorise potential answers or responses, but be prepared to expand on things you have mentioned in the personal statement.
- Important preparation includes rereading your UCAS personal statement. Never include anything in your UCAS application that you are not prepared to speak about at greater length or in more detail at the interview.

- Questions may well be asked on your extracurricular activities. The interviewer may do so either to put you at your ease or to find out about the sort of personal qualities you possess; therefore your answers should be thorough and enthusiastic.
- At the end of the interview, you may be asked if you have any questions. Often, this is simply a polite way of ending the interview, so do not feel that you need to ask anything. Just say 'Thank you, but all my questions were answered during the introductory lecture today and by the students who showed us around. If I think of anything I will contact the admissions department.'
- However, it can also sometimes work in your favour if you do ask a question, as long as it is not something that you could find out for yourself. Try and tailor any questions around the subject itself, showing your genuine enthusiasm for the subject, or, maybe pick up on something you were not able to answer during your interview. This can be another way of demonstrating your willingness to learn.
- Try and explain a topic to a friend or family member who knows nothing about the subject. This is an exercise in communication skills and your ability to clearly exemplify your point.
- Finally, smile, thank them and shake hands. Above all, convey your enthusiasm so that they will remember you at the end of a long day of interviews.

'Within the Department of EEE, we actively encourage all applicants that are expected to meet our very high entry standards to attend their interview afternoon. The reason is not to test their academic ability, but to ensure that they have what it takes to succeed at Imperial in their studies. Perhaps more importantly, we want our high-calibre students to be happy during their stay at Imperial and so it is important to know that they want to study EEE for the right reasons. To this end, during their interview, we are looking for a student who is 'switched on' (as opposed to an applicant having been primed for the occasion) and genuinely keen to pursue a career in EEE (it is not uncommon for parents to push their children into this subject). We ask each applicant a few technical questions, of increasing difficulty, to gauge their limits of mathematics and physics; not necessarily related to EEE. In addition, we are looking for a confident, inquisitive mind, as well as signs of weakness areas. Since places are limited, we try to find signs that the applicant will have a long-term commitment to academic life, as well as indicators that suggest that they will help to enrich the lives of those around them.'

Dr Stepan Lucyszyn, Undergraduate Admissions Tutor,
Department of Electrical and Electronic Engineering (EEE),
Imperial College London

Preparing for an interview

If you are called to interview, it is advisable that you reread your personal statement to remind yourself of what you wrote, so that you can be ready to answer questions on it at interview. Preparation for an interview should be also an intensification of the work you are already doing outside class for your A level courses. Interviewers will be looking for evidence of an academic interest and commitment that extends beyond the classroom. They will also be looking for an ability to apply the theories and methods that you have been learning in your A level courses to the real world.

Essentially, the interview is a chance for you to demonstrate knowledge of, commitment to and enthusiasm for engineering. The only way to do this is by trying to be as well informed as you can be. Interviewers will want to know your reasons for wishing to study engineering and the best way to demonstrate this is with examples of things you have seen, read about or researched. Later in the chapter there is a section on current issues that you can use to kick-start your reading.

Newspapers and magazines

Before your interview it is vital that you are aware of current affairs that relate to the course for which you are being interviewed. *New Scientist* will give you a good grasp of scientific and engineering developments, as will reading the science sections of the broadsheet newspapers. You should also keep up to date with current affairs in general.

Magazines can be an important source of comment on current issues and deeper analysis. There are many specialist engineering publications, such as *The Engineer*, *New Engineer*, *Structure Magazine* and *Aviation Week*. Further details can be found in Chapter 11.

Television and radio

It is also important to watch or listen to the news every day, again paying particular attention to news about scientific and engineering issues. Documentaries and programmes about engineering projects can be enormously helpful in showing how what you are studying is applied to actual situations and events. Keep an eye on the television schedules for programmes or series on anything related to your field of interest, which could range from those aimed at a wide audience (*Grand Designs*, *James May: The Reassembler*) to more factual programmes such as BBC's *Horizon*, *Megastructures*, *What the Victorians Did For Us* and programmes on engineering projects on the National Geographic Channel and Discovery Channel. BBC Radio 4's series *Inside Science*, *Frontiers* and *Material World* are also very useful.

The internet

A wealth of easily accessible, continually updated and useful information is, it goes without saying, available on the internet. Given the ease with which information can be accessed, there is really no excuse for not being able to keep up to date with relevant current issues. iTunes U gives free access to thousands of lectures and presentations from universities around the world; newspapers can be read online ... the list is endless. In this age of information overload, anyone who is serious about keeping abreast of current issues (or wants to be seen as being serious) has unlimited opportunities to do so. Thus, an interviewer is not going to be impressed with a student who claims that he or she has been too busy to know what is happening in his or her chosen areas of interest.

- Check online news websites every day to read the latest news stories.
- If you cannot buy a newspaper every day, look at an online version, for example www.theguardian.com.

Podcasts

Try and listen to engineering- and science-related podcasts in order to keep up to speed with the latest news and innovations within the field. Series such as *The Prepared* or *How I Built This* are good examples of interesting conversations and ideas for would-be engineers. Radio programmes can also be downloaded as podcasts and listened to at times convenient to you; the BBC's iPlayer gives access to current affairs and documentary programmes. BBC podcasts, which are free, include *Inside Science* and *Material World*.

Examples of your areas of interest

One way to make an interview a success is to illustrate the points you are making with examples. It is also easier to talk about something you know about rather than trying to talk in general terms. And if the examples you use are interesting, the interviewer may well want to talk about them rather than ask you the next question on his or her list. But remember, this will only work if you have done your research beforehand. There is nothing worse in an interview than a conversation along the lines of:

You: One of the things that inspired me to study civil engineering was a journey with my parents up the east coast of England, when we crossed the Humber Bridge, the first suspension bridge I had ever seen at first hand.
Interviewer: I see. Can you tell me something about the reasons for building a suspension bridge there rather than a beam bridge?

You: Sorry, I don't know.

A better answer would have been:

You: The Humber estuary is used for shipping, and because of the width of the estuary, a traditional beam or cantilever structure would not have been able to span the space between the banks. Also, suspension bridges have some flexibility and that area is prone to high winds.

The interviewer might then have gone on to discuss the types of forces that are present in a suspension bridge, and about suitable materials, all of which you would be familiar with because you had anticipated this response and had prepared for it.

Here are some ideas to use as examples to illustrate points you want to make:

- civil engineering: names and construction details of bridges or a new airport
- structural engineering: how the use of new materials helped in the building of a new skyscraper or office block
- mechanical engineering: examples of machines that you can discuss (cars, aircraft, wind turbines, etc.)
- chemical engineering: an industrial chemical process
- biomedical engineering: a medical breakthrough that was developed by engineers (such as computerised tomography (CT) scanners)
- electronic engineering: the background on the development of, for example, solid state memory for computers.

The interview

Interview questions are likely to test your knowledge of engineering projects and developments in the real world, since, unlike some theoretical science subjects, engineering is a practical subject aimed at making the world a better place. It is important that your answers are delivered in appropriate language. You will impress interviewers with fluent use of precise technical terms, and thus detailed knowledge of the definitions of words and phrases used in engineering is essential. Potential electrical engineers need to know the technical and microscopic differences between semiconductors and insulators, and to be able to differentiate clearly between electric charge and current; and if you are interested in materials or civil engineering, you need to use words such as stress, strain, elasticity, strength, toughness and stiffness with their scientific, rather than their everyday, meanings.

One popular question is to ask which topics you have enjoyed studying at school. Be prepared for this by doing some revision so that you are not desperately trying to remember details from things you studied a year ago. Try to talk about something that is closely linked to engineering.

You may be asked about your future career plans. If you are applying for a particular field of engineering then your future area of specialty will be apparent, but you may be applying for a general engineering course, or one where you specialise only in the second year, so be prepared to talk about your plans. You may have ideas about where you want to work, in a big company or possibly overseas. It is a good idea to relate your possible plans to research you have done, or to your work experience. It is also a good idea to demonstrate a knowledge of how you gain Chartered Engineer status (see page 124).

You may be asked questions that appear to want your opinion on a recent development or issue. This type of question is asked to see whether you have been thinking about engineering issues, or whether you have been keeping up to date with current issues. Ultimately, the interviewer is not really interested in your opinion but in your ability to formulate arguments and your interest in the field.

Practice of interview situations, like most other things in life, will make you better prepared, less nervous and more confident. Arrange mock interviews with teachers, friends of your family, or with careers advisers.

Some things that can adversely affect the interview include:

- arriving late and flustered
- being unpresentable or scruffy
- being unable to talk about things mentioned in the personal statement
- not listening to the question carefully before answering
- interrupting the interviewer.

Student experience

'I would say I've always been a very curious person with a penchant for problem solving. I was always trying to fix and create. This carried on throughout my GCSE years, however it was during my A levels when a personal tutor asked if I'd ever considered an engineering degree that I first explored the subject specifically. Upon researching different courses and their content, I became absolutely set on studying Mechanical Engineering at university.

'Studying Mechanical Engineering at Newcastle University gave me a wealth of opportunities and experience and created an ambition in me to learn more. The constant learning and keen interest in the subject allowed me to continue growing as an Engineer and as a person by boosting my confidence and familiarity with the subject. One of the main things I enjoyed was meeting people from all walks of life and from all parts of the world but coming together over something we were all passionate about. I had great tutors who were interesting, knowledgeable and extremely supportive. The entire department was great and were always there to guide and assist with anything myself or the group needed.

'Moving to Newcastle was a big change for me as I'd moved up from London. I was of course anxious about the big move, but in no time at all I realised that I was going to thoroughly enjoy my new home. There is so much to see and do and it seemed like there was a friendly face wherever I turned. It became obvious to me that whether you liked history or music or anything in between my new city had it all.'

Cherelle Bovell, University of Newcastle,
BEng Mechanical Engineering

Student experience

'Regarding interviews for mechanical engineering, I would recommend revising calculus, graph sketching and physics, such as mechanics and electricity. I used a site called www.i-want-to-study-engineering.org which has many questions and solutions for many engineering type interview questions. Also, I found that revising from my school textbook was useful for brushing up on old techniques. My interview experiences were as follows:

'First interview:

'I was given a situation where a hopper was filled with sand and placed on some scales. The task was to draw the graph of the reading the scales would show if you let the sand fall out of the hopper and onto the scales, from the time the first grain of sand started to fall until the last one fell. The next question was about a cube hanging from one of its vertices and the task was to find the shape you would get if you cut horizontally across it from different heights. (No drawing was allowed.)

'Second interview:

'I sat a one-hour paper before the interview and we went over it with the interviewer.

'To start, there were some graphs to draw:

$y=(x+1)/(x-1)$

$y=(x+1)^2/(x-1)$

'Then there were questions about circuits with resistors in parallel and in series. In a circuit with a cell of voltage V, there are two resistors in parallel to each other, which are then connected in series to another pair of resistors which are also in parallel to each other. The task was to find the power dissipated by one of the resistors.'

D.S. Barreto Romero, University College London,
MEng Mechanical Engineering with Business Finance

50 sample interview questions

A. About you

1. What first started your interest in engineering?
2. Why do you want to be an engineer?
3. What qualities does it take to be a successful engineer?
4. What have you done to investigate engineering as a course?
5. What have you done to investigate engineering as a career?
6. What field of engineering particularly interests you?
7. What did you learn from your work experience at 'X Enterprise'?

B. Your research into engineering

8. As an engineer you will be involved in creating commercially viable products. Can you discuss a situation where the commercial pressures are in conflict with environmental issues?
9. Give me a very brief outline of the key engineering developments of the twentieth/twenty-first century.
10. What do you consider to be the most significant engineering project in history?
11. Do you have an engineering hero/heroine?
12. What do you think is the difference between science and engineering?

C. Academic questions

13. How does the structure of a metal determine its properties?
14. Why does the molecular structure of wood make it suitable for some building projects but not others?
15. What is the difference between a 'tough' material and a 'strong' material?
16. What do we mean by potential difference?
17. What is the difference between charge and current?
18. Why are the concrete girders used to construct buildings 'T'-shaped in cross-section?
19. How does the internet work?

20. What is meant by conservation of energy?
21. What is meant by a 'cantilever'? How do the stresses on a bridge using cantilevers differ from a bridge using a beam to span the distance between two supports?
22. Which parts of the electromagnetic spectrum can humans detect?
23. Can you explain what is meant by 'proof by induction'?
24. Can a scientific theory ever be proved?
25. What is a semiconductor?
26. What is a machine?
27. What is a computer?
28. What is a bit and what is a byte?
29. What is 27 in binary? What would it be in a system based on the number 4 rather than 2? Or in a number system based on the number 9?
30. Why is nanotechnology so called?
31. How does a car engine work?
32. Why is the distance that an electric vehicle can travel so small compared with a petrol-fuelled vehicle?
33. How does an aircraft stay in the air when it is more dense than air?
34. What is meant by the words 'digital' and 'analogue' when describing communication systems such as TV signals?
35. Why does a bicycle have gears?
36. What limits the maximum height of a proposed new office development?
37. What is a robot?
38. An architect designs a 50-storey office building. How might an engineer test whether it is safe to build it?
39. Can you show the forces acting on a ladder?
40. A designer creates a model for a new type of passenger aircraft. How might an engineer test whether it is safe to build it?
41. What is biotechnology?

D. Unexpected questions (designed to test your ability to think creatively)

42. Murphy's law says that whatever can go wrong will go wrong; for example, if you drop a piece of bread that has jam on one side on the floor, it will always fall with the jam side down. How would you go about verifying Murphy's law?
43. Engineers cause more problems than they solve: do you agree?
44. Is there any such thing as a 'new' invention?
45. People describe new inventions as being the most significant since 'the invention of the wheel'. How do you think the wheel was invented?

46. People describe a good idea as being 'the best thing since sliced bread' – what are the advantages and disadvantages of sliced bread?
47. What do you think the future of Formula One looks like for an engineer?
48. What will replace fossil fuels when they run out?
49. What do we mean by 'alternative' energy sources?
50. Describe this (showing the student an everyday object – a chair, a frying pan, a light bulb, a shoe, a watch) from an engineering perspective.

How to answer interview questions

Introductory questions

Why have you chosen to apply here?

The interviewer will need to be reassured that you have done your research, and that you are applying to the university for the right reasons, rather than because your friend tells you that the social life at that particular university is excellent.

Your answer should, if possible, include the following points:

- first-hand knowledge of the university, for example you came to an open day or you have spoken to students who have studied there; if you cannot visit the university, then at least try to discuss the institution with current or ex-students (many university websites have links to current students who can answer your questions directly)
- detailed knowledge of the course and why it is attractive to you, or how it links to your future career plans. The course might, for example, offer the chance to learn a language as one of the options in year 2, and you could mention this as being something that will help you to work overseas. Or it might offer work placements or the chance to spend a period of time at an overseas university.

Why do you want to be an engineer?

(This question, or a variant on it – What have you done to investigate engineering? When did you decide that engineering was the right course for you? – will almost certainly be asked. It would be considered by the interviewers to be a gentle introduction to the interview because they will assume that you have thought about this, and anticipated it being asked.)

Your answer should include an indication of how your interest started (for example taking apart a radio, building a model out of Lego, something you were taught in a science lesson) and lead on to things you have done to investigate engineering. This would, ideally, involve work experience or an engineering lecture you went to. You could end up by talking about a particular area of interest (mechanical engineering, civil engineering) or a project that interests you (a building, a machine?) and possible plans for your future career.

General questions about engineering

What qualities should an engineer possess?

(Variants on this question might include 'From your work experience, what did you learn about what it takes to be a successful engineer?')

Points you might raise could include: mathematical ability, logic, analytical and problem-solving skills and curiosity. But it is important to expand on these rather than simply list them. Explain, ideally using an example to illustrate what you are saying, why you think that this quality is important. Examples can be drawn from your work experience, your wider reading, or a lecture.

Here is an example:

> 'The ability to solve problems is very important. I really became aware of this when I was doing my work experience at a local engineering company. They were making low-voltage lamps for use in recessed lighting fittings in houses and offices, but in one particular building, the lamps kept blowing. In the end, one of the engineers decided that the problem must have been in the transformer rather than the lamp itself, and so he looked at where the transformers for each lamp were situated. It turned out that they were short-circuiting because the cavity above the false ceiling was damp.'

What does a chemical engineer do?

(Variants on this could include asking for definitions of engineering, science or technology.)

Again, try to illustrate your answer with an example or the details of a conversation that you had with an engineer:

> 'Chemical engineers primarily design chemical manufacturing processes and then find development initiatives to enable those schemes. I was told on my work experience with a chemical engineering firm, that in order to be successful as a chemical engineer, you need to have a proficient knowledge of the principles of all three sciences and mathematics in order to address problems such as the production or use of chemicals, the development of fuels, food, drugs, etc. It is not about being left- or right-brained, to be creative – as you must be – you need to be using both.'

Questions designed to assess your clarity of thought

You may be given an open-ended question about something you have already studied. The point of this type of question is not so much to test your knowledge or academic level (because this will be clear in the grade predictions and exam results on the UCAS application) but to see if you can think logically and in a structured way. So, your answer should really be an exercise in 'thinking aloud'; that is, talking the interviewer through the steps to your final answer. An example is given below.

Why are metals so useful to engineers?

You could start from first principles by describing the microscopic structure of a metal. This shows that you can approach problems in a logical way while also giving you some time to think about where your answer is going: 'A metallic structure consists of a lattice of positive ions surrounded by a "sea" of delocalised electrons. It is this structure that gives metals their useful properties.'

You might then go on to look at a number of properties in detail: 'The most obvious properties this gives metals are good electrical and thermal conductivity. Electrical conduction is through the flow of electrons through the lattice, and since they are not attached to any particular atom, they can move freely. Metals are also good conductors of heat because the electrons are able to transfer energy as they move in addition to the vibrations of the lattice.'

You could then move on to other properties that make metals so useful, describing each one in turn. You would probably include malleability, and the use of physical processes to alter the strength, stiffness or toughness of a metal to suit its intended usage.

Questions that assess your ability to analyse or to solve problems

You may well be confronted by a question about a situation that you have not covered in your studies. Don't worry. The interviewer will know that this is a new area for you. What he or she is looking for is not for you to immediately give them the correct answer, but rather how you can take things that you know and apply them to new situations.

Applicants for one university were asked 'What percentage of the world's water is in one cow?' Of course, no one (including the person who asked the question) knows the answer to this. What they were interested in, as discussed in the previous example, was in the candidate's ability to approach a problem from first principles and to arrive at an answer in a logical and structured way. So, an answer of 'I don't know' would not be very useful to your chances of being offered a place.

A better answer might start with: 'Well, I suppose I might begin with trying to estimate how much water there is on Earth. I know from my physics what the radius of the Earth is, and so I could work out its surface area. I could then make an assumption about the average depth of the oceans and work out their volume ...' Rather than listen in silence, it is likely that the interviewer will help you by giving you hints or guiding you. But they can only do this if you explain every step.

When I talk to students about what they worry about when they are preparing for their interviews, they always say, 'What if I cannot answer a question?' And here is what I say to them.

- Interviewers are aware of the level you have studied to, and so will have a good idea of what you should know and may not know.
- Therefore, it is likely that any 'new' topic that you are confronted with at the interview has been asked to see how you think rather than what you know.
- Approach all questions from first principles, for example GCSE knowledge, and then build up your answer.
- If your answer requires you to draw a sketch or do a calculation, ask if you can use a piece of paper or the whiteboard in the interview room.
- Don't be afraid to ask for help, but do this by asking for comments on what you think is the right approach: 'I think I would start by looking at the forces on the body – is this right?'

Questions to show your interest in engineering

Anyone can say that they are interested in engineering, but by applying to study engineering at university, you are embarking not just on a short period of study but on your future career as well. An interviewer (who is almost certainly an engineer) will want to be reassured that you are serious enough about the profession to keep up to date with developments and engineering issues. So, questions such as 'Tell me about an engineering issue that you have read about recently' are designed to see if you keep abreast of current events. How do you ensure that you are prepared for such questions?

- Watch the television or listen to the radio news on a daily basis. Read the quality newspapers as often as you can, and keep a scrapbook of engineering-related stories.
- Check websites for engineering stories. A good starting point is the BBC website (www.bbc.com), which has a section on science and technology.
- Visit the news sections of the websites of the engineering institutions and professional bodies (see Chapter 11 for website addresses).
- Talk to engineers.
- If you can, go to public lectures at universities. Details can be found on the university websites.

- Watch podcasts of university lectures on iTunes U.
- Download or listen to radio programmes such as BBC Radio 4's *Material World*.

Steering the interview

There will be issues that you want to raise in the interview, things that will demonstrate your research, commitment and personal qualities. Rather than walking out of the interview disappointed that you did not have the opportunity to discuss these things, try to bring them into the conversation. For example, you may have been to a lecture on developments within the electronics industry at a local university one evening and you want to talk about this. There are likely to be many ways that you can do this. You might be asked why you want to be an engineer, and during your answer you could say 'and the thing that really convinced me that electronic engineering was the right career for me was listening to Professor Smith talking about nanotechnology at a lecture that I attended at Surrey University last month'.

In all probability, the interviewer will then ask you more about this, and you can then talk about something that you know about, rather than having to face questions on a topic with which you are less familiar.

Before you go to the interview, write down a list of things you want to talk about, and think of ways that you may be able to do so.

Current issues

As a potential university engineering student you need to demonstrate your interest by keeping up to date with current issues and developments. Engineering is an ever-evolving subject, with new materials, processes and products being developed every day. Just think about the rapid changes in the field of communications over the past 15 years, for example. Once you have identified your particular areas of interest, you need to keep researching and reading about them, perhaps keeping a scrapbook (either a physical one or on your computer) of news articles that are relevant to your chosen area of study, and using this as part of your preparation for an interview.

In this section, I have covered some broad topic areas that should be familiar to potential engineers. The summaries in this chapter are included to give you an illustration of the kind of events and other news items you should be reading about. In other words, they are a starting point for your own research, rather than being an easily accessible source of information to memorise prior to your interviews.

Engineering and materials

Engineering is about making practical things and so engineering projects require materials – that is obvious! But what factors determine which materials are suitable for a particular project? And how do engineers go about choosing between materials that may appear to have similar properties? To do this, engineers need to be able to describe the properties of the materials they are looking for. Physicists and engineers have to be very precise in the way they describe the properties of materials, and words that we use in a very general sense in our daily lives have very specific meanings for engineers. Words such as 'tough', and 'strong'; or 'elastic' and 'plastic'; words that we use in normal conversation and might appear to mean similar things have precise definitions when used in an engineering context. Prospective engineers need to learn the vocabulary of materials.

When we talk about properties, we mean how a particular material behaves when subjected to external factors – light, forces, heat, electricity, time. And a property that might be useful in one context, for example, stretching elastically, might be problematic in a different situation.

Some of the properties of materials that engineers are interested are:

- electrical conductivity
- thermal conductivity
- how they behave when subjected to forces
- how their behaviour changes over time
- resistance to corrosion
- how they are affected by heat or temperature
- how different materials can be joined
- resistance to chemicals or changing weather conditions
- toxicity
- aesthetic properties – will people want to buy or use the product?

Consider your smartphone. What materials are used and why? Electronic components rely on their ability to conduct (or not) electricity. The electronic circuits and microprocessors in your phone are built out of semiconductors. Semiconductors are somewhere in between electrical conductors and electrical insulators and the way that they conduct electricity depends on the materials used. And so electronic engineers are interested in the properties of semiconductors. The flow of electricity through your phone through the conducting parts of the circuits generates heat, and so the engineer is also concerned with minimising this and thus needs to understand the different metals that are available. Some of these are cheap, some expensive, some last a long time, some corrode, some are safe to be in contact with, some are toxic.

Your phone relies on a battery to supply the electricity that it needs to function. This battery contains chemicals. It needs to be able to be recharged regularly without deterioration of the chemicals, and to have a long lifetime. And what happens when the phone is defunct and has to be thrown away? What effect will these chemicals have on the environment? How safe are the batteries? If you were the engineer that developed the batteries in a recently available smartphone that appeared to have a tendency to catch fire and the company that developed it had to recall millions that had been sold, your future job prospects might not be that promising.

What does the designer want the phone to look like? Should it be available in different colours, or look metallic? And how resistant will it be to wear and tear or accidents? Some materials are shiny, some are transparent. Some are strong, some will crack when the phone is dropped. Some get hot very quickly. Materials engineering is a field that answers questions like these.

Let's look at another example: building a new iconic skyscraper. An architect will create the initial design of the building, which will be based on aesthetic ideas, as well as thinking about what the building will be used for – 'form and function'. This would probably start as a rough sketch and then the architectural practice will work on more detailed designs. This is where the structural engineers will be involved. First and foremost, the building has to be structurally viable; that is, it shouldn't fall down. So the structural engineer will need to look at all aspects of the structure: will it be built around a central concrete core with the exterior fixed to this, or are the external walls going to be load-bearing? Concrete is very strong when subjected to compressive forces, but cracks easily when tensile (stretching) forces are applied. So if tensile forces are involved, then it has to be strengthened or toughened (see below) by adding other materials to reinforce it or pre-stress it. If the building is going to be tall, then how flexible does it need to be in strong winds? How will people get to the upper floors? What materials should be used for the lift cables? What about ventilation? What type of glass? A building with a glass exterior will get very hot because of the greenhouse effect.

A brief explanation of some of the commonly used terms that describe materials or their properties is given below. But you should read more about the properties of the materials that are used in the particular field of engineering that interests you.

Types of materials

Metals: the atomic structure of a metal is what gives metals their most useful properties. Because the electrons are relatively free to move through the fixed crystal lattice structure, metals are good conductors of electricity and heat. These 'delocalised' electrons also make metals malleable and shiny.

Polymers: made up from long chains of repeating units. These molecular chains can be arranged in more or less orderly patterns, or tangled up randomly. This means that polymers can usually be stretched and can be moulded into shapes. Polymers are chosen for particular uses by looking at their elastic or plastic behaviour, and by looking at how they react to heat, light or chemical exposure. In October 2018, the introduction of self-healing materials on an industrial scale has moved ever closer. Chemical engineers from the Massachusetts Institute of Technology (MIT) designed a polymer that is able to react with CO_2 from the air, which would then grow, strengthen and potentially even be able to repair itself. It is designed as a way of positively using greenhouse gases and researchers claim that it could, in the future, be used for construction material or a protective base layer.

Amorphous: as the name suggests, the atoms and molecules are randomly arranged with no long range discernible order. Glass and ceramic materials are amorphous in structure.

Composites: composites are composed of different materials (or the same material in different orientations) and exhibit the best properties of each. Reinforced concrete incorporates steel rods within the structure so is strong in compression and also in tension.

Properties of materials

Strength: strength is a measure of how much stress a material can withstand (a similar measure to pressure) before breaking.

Toughness: not the same as strength. A tough material might or might not be strong, but it will not be brittle, that is, it won't break through cracks appearing and spreading through the material. So glass can be strong but is not tough (unless it is laminated or treated to improve its toughness), whilst a digestive biscuit is neither strong nor tough. A polythene bag might not be very strong but it is tough.

Stiffness: a measure of how easily the material deforms – bends or stretches or compresses – when forces are applied.

Brittleness: brittle materials tend not to stretch very much when a force is applied, but then suddenly break. Glass is a brittle material.

Elasticity: elastic materials deform when forces are applied, but then return to their original shape and size when the force is removed. Why is an elastic band so named? Because it exhibits elastic behaviour.

Plastic behaviour: the opposite of elastic. If the force is removed the material does not return to its original shape. Take a plastic supermarket bag and cut a thin strip from it and then stretch it. Initially it will not stretch much but when you apply a certain force it will suddenly be easy to stretch by a significant amount. But when you stop applying the force it will not go back to its original size. So a plastic bag is so called because it behaves plastically.

Malleability: metals are malleable. It means they can be hammered or pressed into new shapes.

Electrical conductivity: when a potential difference is applied to a conductor, how much current will flow? The higher the current, the higher the conductivity. Electrical conductivity may be a desirable or undesirable property. Electrical insulators are also required in any devices that use electricity. Semiconductors are materials that conduct electricity by an amount determined by how it is made or at what temperature it operates.

Thermal conductivity: a measure of how much heat flows in a material. As with electrical conductivity, both good and poor thermal conductors are useful to engineers. Cooling is important in many mechanical devices where heat is generated, and so good conductors are needed to remove the heat efficiently. But the handle of your kettle should be a good thermal insulator so you do not burn your hand.

Failure mechanisms: a material or component can fracture or stop being suitable for its chosen purpose in many different ways and from many different causes. They can crack, deform, melt, corrode, decay, suffer from metal fatigue, or stop being useful in many other ways. The causes of this could be defects in the molecular structure (metal fatigue is caused by dislocations within the structure), the application of too much force or heat or electric current, through chemical reactions, surface scratches or even by being joined to other materials using unsuitable adhesives or techniques.

Commercial issues

Someone has to pay for the things that engineers produce. Sometimes you pay through your taxes (roads, bridges, medical equipment), but in most cases the things that engineers make have to be useful or attractive enough for people or companies to pay for them. Two very similar, in terms of functionality, smartphones or cars may vary in price by a factor of four or five simply because of the materials used.

Engineering, ethics and the environment

Engineers are professionals and, as such, they have to behave in a professional manner, and ensure that they act ethically at all times. Engineers need to be trusted by the public. The work that engineers do affects millions of people – think of engineering projects that involve defence, power generation or the provision of clean water, for example.

The Royal Academy of Engineering and the Engineering Council sum up the obligations of engineers in their Statement of Ethical Principles:

1. Honesty and integrity
2. Respect for life, law and public good

3. Accuracy and rigour
4. Leadership and communication

Source: www.raeng.org.uk/policy/engineering-ethics/ethics#statement.
Courtesy of the Royal Academy of Engineering.

In the publication 'Engineering Ethics in Practice' there are case studies that outline how these principles apply to real-life situations. They illustrate how, for example, engineers need to get the right balance between the economic demands of the companies that they work for and ethics. A recent example of this is the problems that Volkswagen have had with software that affected the readings for emissions from some of their diesel vehicles. The big topic of 2020 was the 5G network and the engineering of that balanced against the ethical implications of it. Other ethical issues involve surveillance, cutting costs to ensure that projects finish on budget, or the use of poor quality or dangerous materials.

Source: www.raeng.org.uk/publications/other/engineering-ethics-in-practice-full.
Courtesy of the Royal Academy of Engineering.

The environment

Almost everything that engineers produce, design or operate has a potential impact on the environment; they are at the forefront of the global climate crisis and it will be them – alongside the collective will – that will resolve it. Some large-scale projects have an obvious effect – a new dam, wind farms, new housing or office developments. Cars and aeroplanes produce emissions as a result of converting fuel into kinetic energy. Power stations produce electricity from fossil fuels or nuclear reactions and the by-products of these can be harmful to living things or have long-term effects on climate. Discarded batteries and electronic devices contain poisonous chemicals. Engineers are increasingly having to incorporate safeguards into their designs as public pressure is put on them to reduce environmental damage and avoid the depletion of natural resources and habitats.

Green buildings

It is estimated that between 30% and 40% of energy use in developed countries is associated with buildings, most of which is used in heating, cooling and electrical systems. The concept of a 'green building' is a relatively new one. By describing a building as being 'green', we mean that it is designed to be as energy efficient as possible. This can be achieved by:

*   using building materials that do not require large amounts of energy to produce – plastics, concrete and metals require significant amounts of energy, whereas natural materials such as wood require much less
*   insulating buildings effectively to avoid heat losses

- designing buildings with natural ventilation, such as using natural convection to produce cooling, rather than relying on air conditioning
- aligning buildings so that sunlight can be used either as a source of heating or of electricity in photovoltaic cells
- using 'thermal mass' to absorb heat and then release it naturally
- re-using water through recycling facilities.

Aerospace

Extreme weather in space has now been upgraded to the National Risk Assessment (NRA). The Royal Academy of Engineering is in consultation with relevant partners over the establishment of infrastructure that is able to withstand solar storms. For example, Global Navigation Satellite Systems (GNSS), and developing GNSS's capacity to operate with holdover technology for up to three days. Extreme space weather is a clear and present concern, and this is likely to be the focus of certain strands of engineering from now on.

Hybrid vehicles

Hybrid vehicles are vehicles – cars, trains, trams – that have two separate sources of energy. While the idea is not new (the moped is a hybrid vehicle as it is powered by a motor or engine and by the rider), car manufacturers are now devoting enormous resources to new ranges of hybrid electric vehicles. Hybrid petroleum electric vehicles are powered by internal combustion engines that run on petrol, diesel or gas. They also contain heavy-duty batteries that are charged as the engine is running, and the car can be powered electrically during parts of its journey. Other hybrid vehicles use the engine to compress air or hydraulic fluid, which can then be used to drive the wheels, and there are trams that are able to use a combination of diesel fuel and overhead electric cables. What all of these have in common is that they reduce the use of fossil fuels (and so reduce greenhouse gases and pollution) and utilise energy that might otherwise be wasted when the car is running but not moving.

Other examples of how engineers are adapting to the needs of the environment include:

- the development of new energy-efficient materials, such as composites that make aircraft bodies lighter, or insulators that prevent heat losses
- electrical conductors that have low electrical resistance in order to reduce heating when conducting electricity
- stronger, lighter materials for electrical devices
- ways of extracting minerals or fuels from below the Earth's surface that have lower environmental impact
- new ways of producing energy that do not rely on fossil fuels (see below).

Smart motorways

In 2006, the government introduced smart motorways in order to control the flow of traffic and the safety of the roads. Highways England reports that congestion costs about £2 billion every year. A further 200 miles of road are currently planned to be upgraded to become smart motorways, which represents a further extension of this enormous civil engineering infrastructure project.

Plastic waste

The ever-increasing amount of plastic in the world's oceans is a clear and present problem. Engineers are looking at ways of combating this issue. Innovations such as Seabins – which drag in, on average, over 4kg of debris per day – sea walls, and water wheels in rivers to catch debris have all been trialled. The key is a global, unified effort: engineers everywhere need to work as a team to combat a problem that continues to escalate. Organisations such as Ocean CleanUp are at the forefront of this and you should research them before your interview.

Engineering and energy

Engineers use energy to create their products, and to run them. Engineers also create the structures and processes that produce usable energy. Engineers, therefore, are at the forefront of the search to develop new energy sources. As well as being finite sources of energy (in other words, they will one day run out, or, at the very least, become uneconomical to extract), fossil fuels cause environmental damage and are closely linked to climate change and global warming. Many fields of engineering have an interest in developing alternative or renewable energy sources. While there are no set definitions of these words, they generally refer to energy sources that do not involve fossil fuels and/or those whose consumption does not deplete the planet's natural resources or reserves. The engineers who are primarily involved in developing alternative energy sources are electrical engineers, electronic engineers, bio-engineers, mechanical engineers, geo-engineers and chemical engineers.

There are a number of ways of categorising energy sources (see Tables 2 and 3 on the following page).

Alternative energy sources that are currently used commercially are:

- geothermal energy
- hydroelectric power
- nuclear fission
- solar power (solar panels to produce heat and photovoltaic cells that create electricity from sunlight)
- tidal power

Table 2 Categorisation of energy sources based on origin

Energy sources derived from the sun	Energy sources not derived from the sun
Solar	Geothermal
Wind (heat from the sun creates areas of low and high pressure, causing wind)	Nuclear fusion
	Nuclear fission
Wave (winds cause waves in the sea)	Tidal
Fossil fuels (coal, oil, gas, etc. were once living things)	
Hydroelectric (water sources, such as rivers, were created by precipitation of rain, caused by the sun's heating)	
Biomass	

Table 3 Categorisation of energy sources based on effect on source

Renewable*	Non-renewable
Solar	Fossil fuels
Wind	Nuclear fission
Wave	
Tidal	
Hydroelectric	
Geothermal	
Biomass (since although the individual sources cannot be used again once the energy is extracted, the supply can be maintained by, for example, planting more oil palms)	
Nuclear fusion (considered renewable since the likely source would be sea water, which is effectively inexhaustible)	

* By 'renewable' we generally mean that the source of the energy is unaffected by the extraction of energy. For example, every time coal is burned there is less coal remaining on the Earth, whereas if a wind turbine creates electricity it does not affect the source (movement of air, which is caused by solar heating).

- biomass (using bacteria to produce gas through the digestion of organic materials, making fuels out of natural materials such as palm oil)
- wave power
- wind power.

An alternative energy source that is in development is nuclear fusion.

Recently, fracking has become one of the most contentious new ways of sourcing fuel, causing split opinion across communities. It is worthwhile understanding the arguments for and against this process, while considering it from an engineer's perspective, i.e. the hydraulic process involved in splitting the rocks and the repercussions of doing so.

Engineering and the NHS

Right now, the NHS needs innovation. The only way it can continue to develop and aid the lives of others is to have the technology to support it. With the day-to-day pressures faced by the NHS, the engineering profession is finding new ways to support the needs of the health service for the benefit of those working in it and those using it. Much like the healthcare system, engineering is systems-based and is a sector always looking for innovation. It has long been acknowledged by those within both sectors that cross-disciplinary learning could yield new understanding and greater efficiency.

Engineering in the NHS could mean anything from the regeneration of ageing hospitals to inventing new medical technology. Biomedical engineering is now an established degree course.

In October 2018, King's College London announced that it was teaming up with engineering firms Nvidia and their AI Tech to utilise their supercomputer in the fields of radiology and pathology. Given that the number of patients requiring radiology is reportedly increasing around 16% every year yet the number of radiologists required to analyse that data is nowhere near in step, the supercomputer could provide a much faster analysis of imagery. This is just one example of engineering partnerships through technology that can aid the NHS in the long term, but given that medicine is a technology-focused world, you should be able to find many more examples before your interview.

Remember, too, that when we talk about engineering within the medical field, it does not have to be technology-related: for example, prosthetic and artificial limbs are currently one of the most progressive fields of biomedical engineering.

The commercial space race

Elon Musk or Sir Richard Branson? Currently, Forbes has the USA ahead of Europe in the race to send a commercial flight into space. In 2018, Tesla sent a car into space on a privately made rocket. In 2020 SpaceX announced that it had developed the most powerful rocket ever: Falcon Heavy; Elon Musk has a grand ambition of reaching Mars by 2022. What are your views on that? Anything is possible, it is just a question of when and how commonplace it will become. Commercialisation is a term often used in this conversation, but remember that many companies share common goals and due to the funds needed to achieve such goals, we will not be looking at an overcrowded space channel for at least the immediate future.

Aviation and aeronautical engineers are at the forefront of pioneering these passenger spacecraft. Their design and structure are critical as there are different factors to consider when operating for a commercial audience and a great variety of engineering disciplines will be utilised in the process.

Engineering business case histories

Try to keep up to date with stories associated with engineering businesses and entrepreneurs. Most new engineering projects and developments are driven by commercial companies that want to sell their products or processes, and while some new devices are developed by university research departments, these are often funded by industry.

The following list aims to provide you with a starting point for your research, as there will undoubtedly be new names that will become more relevant by the time you read this book:

- Apple's iPad, iPod and iPhone
- the Android, iOS and Windows operating systems for mobile devices
- Google
- apps
- 5G networks
- inventors, scientists and entrepreneurs, such as Elon Musk or Tim Berners-Lee
- SpaceX or SpaceShipTwo
- Virgin Galactic
- multinational construction companies, such as Arup
- aeronautical engineering companies, such as BAE Systems
- UK tech companies (for example, those based in 'Silicon Fen', such as ARM)
- Boeing 787 Dreamliner – note the Boeing 737 has been grounded
- Airbus A220
- biotech companies, such as Gilead Sciences
- oil and petroleum companies that are developing alternative or renewable sources of energy, such as Statoil
- online school platforms as educational engineering to deliver innovative education
- 4D cinematic technology
- digital money
- hyper-personalised drugs

- alternative energy schemes, such as the proposed tidal lagoon in Swansea Bay
- building tall structures, such as the Burj Khalifa in Dubai or the Shard in London
- Battersea Power Station and the change of ownership
- Qatar Olympic building programme
- issues surrounding the excavation and distribution of rare earth metals
- collapse of Carillion and the knock-on effect on engineering firms such as Vaughan Engineering
- fuel cells that produce electricity from hydrogen
- self-heating polymers
- how engineers deal with natural disasters, such as floods, earthquakes and tsunamis
- environmental issues, such as the safe processing and storage of used fuel rods from nuclear power stations, or polluting metals used in mobile phones or computers.

Did you know ... ?

The word 'engineer' comes from the Latin '*ingenium*', meaning cleverness.

---

7 | Don't get your wires crossed: Non-standard applications

Perhaps you are studying for a mixture of examination qualifications, or you have had a gap in your education. You may have already started a degree course in another discipline and want to change direction. This chapter applies to students who may be applying to university as mature students, perhaps with qualifications other than A levels or the equivalent, to international students who are applying from outside the UK and to those with disabilities.

Whatever your situation, the first thing you should do is make contact with some universities (either by telephone or via the email addresses given on the university websites) to explain your situation and ask for advice.

Changing direction

Not everyone at the age of 17 has a clear idea of where their future lies. Often, university or career choices are made on a whim, on the advice of parents, teachers or peers; or simply through lack of experience of the chosen discipline. People change, and luckily universities understand this and are used to dealing with it. Making the wrong university choice should not lock you into a study path or career that is unsuitable, unviable or unpleasant. The case study below illustrates this.

> **Student experience**
>
> 'I had always wanted to study engineering but towards the end of sixth form I began to enjoy my biology lessons more than my physics lessons and therefore made a last-minute decision to take up biology with the added practicality of forensics at university. Unfortunately, I did not enjoy it as much as I had hoped since it

contained very little of my favourite and strongest subject, mathematics. At the end of my first year at university I decided to change from forensic biology back to my original choice of mechanical engineering, and it was very easy to do so. Every student is allocated a personal tutor to provide support and answer any queries we may have. I had a very brief meeting with my personal tutor, who also happened to be my course director. She advised me to do a bit of research about which course I wished to change to and, if I was completely happy, then the next step would be to collect a "course transfer" form from the faculty office. The form then simply needed to be signed by my current course director, my new course director and myself. I emailed the course director for the school of engineering and arranged a meeting; he reviewed my UCAS application for forensics and helped me to decide which course would be best for me. We discussed my options and the course structure for all different branches of engineering. I felt very confident with my choice and the necessary documents were signed in the meeting and within a few days I received a confirmation email about the change.'

Aliya Foster, Kingston University, MEng Mechanical Engineering

Mature students

Mature students (defined as students who are over 21 at the start of their proposed courses) are usually:

- applying with appropriate qualifications, for example A levels, but have not been to university and are now applying after a gap of a few years, or
- applying for a second degree, having graduated in a different subject, or
- applying with no A levels or equivalent qualifications.

If you come under either the first or second of these categories, you apply using the same route as first-time applicants. However, you should contact the universities directly to discuss your situation with them, and to get their advice. The structure of your personal statement will need to be different from that produced by a student who is still at school or college. It should include:

- a brief summary of why you are applying now and what you have been doing since you completed school or college education
- your reasons for the change in direction
- an explanation of any gaps in your education or work history

- a discussion of any appropriate skills or experiences gained in your previous jobs or degree studies.

Most universities encourage mature students who want to apply for entry to degree courses, taking into account their work experience and commitment as part of the entry criteria. Mature students have often left school without the appropriate academic qualifications for university entrance in order to start jobs or careers, in which case there are many Access courses in colleges around the country that specifically prepare mature students for higher education. Others may have studied at degree level in another, non-related, discipline.

Universities often encourage mature students to apply because they:

- bring valuable real-life experience to the faculty
- can be more mature in their study methods than school-leavers
- have had more time to think about what they really want to study
- have a better understanding of the links between study and work.

The best way to find out about acceptable Access courses is to contact the universities directly and ask which ones they recognise or recommend. The engineering institutes (see Chapter 11 for a list of these) provide details of schemes that allow students who have studied on apprenticeship programmes to progress onto a Higher National Diploma or degree course.

If you are applying for a degree course as a mature student without using the Access course route, you should:

- use the UCAS application system as described earlier in this book
- select 'no' to the question 'are you applying through your school or college' when prompted on the online application
- fill in the section on employment as comprehensively as possible, ensuring that there are no periods of time since you left school that are unaccounted for
- ask someone suitable (your current employer or a previous teacher) to act as your UCAS referee
- ensure that he or she knows what the universities require from the referee (you can point them towards the UCAS website, which has a section on information for referees)
- send a more detailed CV directly to the universities once you have received your UCAS number (and quote this on all correspondence with the universities).

It is important to emphasise that the universities are keen to recruit serious, motivated and committed students onto engineering programmes, and mature students tend to fit this description extremely well. You will find that the university engineering departments are eager to help suitable candidates apply and they will be able to provide you with advice and feedback if you contact them prior to applying.

International students

In the 2019 application cycle, UCAS (www.ucas.com) applications and acceptance statistics showed that 1,755 EU students (a significant drop from the 2018 figure of 2,090) and 4,695 non-EU international students gained places on engineering courses.

International students are usually:

- following A level (or equivalent) programmes either in the UK or in their home country, or
- studying for local qualifications that are recognised as being equivalent to A levels, in their own country, or
- studying on academic programmes that are not equivalent to A levels.

Students following A level or equivalent programmes should apply through UCAS in the usual way. All of the information in this book is equally applicable to them.

Students studying qualifications that are accepted in place of A levels can also apply through UCAS in the normal way, from their own country. The UCAS website (www.ucas.com) contains information on the equivalence of non-UK qualifications. These include the Irish Leaving Certificate, European Baccalaureate, and some international O levels and A levels. Information on the equivalence of other qualifications can be found on the UK government's qualifications website (www.enic.org. uk).

Students who do not have UK-recognised qualifications will need to follow a pre-university course before applying for the degree course. These include:

- university foundation courses that are offered by many UK colleges and universities. For example, the NCUK (www.ncuk.ac.uk) programme includes specialist science and engineering foundation paths. Typically, these programmes last nine months and involve 20–25 hours of intensive tuition a week, plus specialist English help
- university foundation courses set up by, or approved by, UK universities or colleges, but taught in the students' home countries
- A level courses (normally two years, but in some cases can be condensed into one year) in schools and colleges in the UK. A levels allow students to apply to all UK universities, including the top-ranked universities such as Oxford, Cambridge and Imperial College.

Foundation (or 'pathway') courses are not accepted by all UK universities. You should check with your preferred universities which courses they accept before committing yourself. Representatives of UK universities, schools and colleges regularly visit many countries around the world to promote their institutions and to give advice. You can also contact the British Council to get help with your application.

International students should note that they are required to pay higher fees than UK students. UK students have their fees capped at a maximum of £9,250 per year, whereas fees for international students on engineering courses are likely to be between £15,000 and £25,000 per year. Accommodation and meals will be extra. How much living costs are depends on where you study, but, as a rough guide, about £1,000 a month should cover food, accommodation, books and some entertainment costs.

Following Britain's departure from the EU in January 2021, EU students will no longer be classed as 'Home' students. EU students starting their courses in September 2021 will not be eligible to apply for student funding.

One of the reasons why international students are, on the face of it, less successful in their applications than UK students is that they, their teachers or their referees are unaware of what is required, particularly if they have experience of applications for universities in other countries. They have also been up against a high volume of EU students though this may no longer be the case, following the removal of EU students' 'Home' status after Brexit. All of the information on personal statements and interviews in this book applies equally to UK and international students.

If you are applying from outside the UK, you must ensure that your referee also understands what he or she needs to write. The UCAS website (www.ucas.com) contains a section giving advice to referees. In general, your referee should address the following points:

- some background information about you – a brief overview of your recent education, and why you are applying for a place at a UK university
- some background information about themselves – what their relationship is to you, how long they have known you, and on what basis are they able to comment on your academic and personal qualities
- an assessment of your academic and personal strengths
- an assessment of your suitability for a course and career in engineering.

English language requirements

Students whose first language is not English will be required to demonstrate that they have the right level of English language to cope with the course. Each university will specify a minimum score on an approved SELT (Secure English Language Test) such as IELTS (International English Language Testing System). Students who have sat IB or IGCSE may not be required to sit an extra English language test. The UCAS Guide for International Students has details of English language requirements: www.ucas.com/undergraduate/applying-university/ucas-undergraduate-international-and-eu-students.

Checklist for international applicants

1. Check that your academic qualifications are accepted by universities. The UCAS Tariff lists all international qualifications that, in theory, are accepted. If your qualifications are not included in the Tariff, contact the universities directly.
2. Check that the level of your qualifications (grades, marks, predicted grades or predicted scores) are at the level required by the universities. The Course Search facility on the UCAS website will give you links to the universities' grade/score requirements.
3. Go to the international pages of the university websites to see if there are any specific language requirements, for example a minimum IELTS score. Some universities will specify an overall score only, for example 7.0. Others may also have particular requirements for each section of the test, such as 7.0 overall with at least 6.5 in each section.
4. Use the international pages to see whether there are any scholarships available to international students from your country.
5. Follow the advice in the engineering sections of the universities' websites regarding any specific information they want to see included in the personal statement.
6. Use the university websites to check the fee and accommodation arrangements.
7. If you are studying at an international school that has sent students to UK universities in previous years, it will be aware of its role in the UCAS application. The school is likely to already be registered with UCAS. If you are studying at a local school that has not sent students to UK universities, you will need to register as a private candidate on the UCAS website, and discuss the reference with someone suitably qualified to write it, perhaps one of your teachers or a previous employer.
8. Be clear about the deadlines for applications.
9. Check whether you need a visa to study in the UK (see below).

Visa requirements

Students from non-EU countries generally need a visa to study at degree level in the UK, although students from some countries are exempted. The UK government website has full details of whether you will need a visa (www.gov.uk/check-uk-visa). The stages in obtaining a visa are set out below.

1. You have been made an offer from an institution that is able to act as a sponsor for your visa application (www.gov.uk/government/publications/register-of-licensed-sponsors-students).
2. You have accepted the offer and have fulfilled the academic and English language requirements.
3. The institution has given you a Confirmation of Acceptance for Studies (CAS).
4. You apply for the visa, presenting the required evidence. Academic evidence such as an IELTS certificate or examination results will be listed on the CAS, and the gov.uk website will list other evidence, such as proof of funds to cover the tuition and living costs.

Note: If you are from the EU, EEA, Liechtenstein or Switzerland and you enter the UK for study after 31 December 2020, you will now require a student visa in order to attend a university in the UK.

Students with disabilities and special educational needs

Universities welcome applications from students who have physical or other disabilities or special educational needs, and they have well-established support systems in place to provide assistance and special facilities. The services offered can help with a wide range of disabilities including sensory (visual/hearing) impairment, mental health difficulties, mobility impairment, dexterity impairment, Asperger's syndrome or other autism spectrum disorders, chronic medical conditions (e.g. diabetes, epilepsy or asthma) and specific learning difficulties (e.g. dyslexia or dyspraxia). In all cases, you should contact the universities directly, before you apply, to explain your particular needs and requirements. They will then be able to give you information on how they can help you. The website www.disabilityrightsuk.org contains useful links and information.

Case study

Duyanh went to the University of Cambridge to study chemical engineering. He has now started his own company in the chemicals sector.

'On completion of my A levels, I chose to study chemical engineering as it was closer allied to my future ambitions in my home country in Vietnam, and this is where I have been since graduating, running my own chemicals company.

'As an international student, I found the application process difficult and was grateful for the support I received. As chances for international students are fewer than for home students, I would advise anyone in a similar situation to get as much help as they can. Ultimately, what works is to make the application personal to you and not to use anyone else's words. I like to think that I got to where I wanted to be as a result of my interest in and engagement with the subject, which I was able to display in interview. You need to be strategic about where you want to apply: don't be fixated on London, for example, where application numbers are higher. You need also to remember that if you are studying A levels, unlike overseas qualifications, you are learning skills necessary for university and it is worthwhile highlighting those skills in your statement.

'Also, remember what makes you unique.

'In my first year, I took the engineering route which was the correct path for me as it gave me a broad overview of engineering. In my second year, I became particularly interested in the module on Process Operations and, specifically, Homogeneous Reactors, which made me want to do further research. This is what I enjoyed most about my course: the freedom to investigate whatever we wished, and to create and to innovate.

'I am now running my own company. The skills that an engineering degree gave me are not just applicable to the scientific value of the field, they are also relevant to how to run a company. It is a systems-based degree after all, and the modelling of a company in that way allows for a greater understanding of how to make an engineering and distribution-based company efficient and productive.

'Engineering is one of the most diverse and rewarding careers that I could have hoped to study, and even though it is arguably harder for an international student, that should just make you want it more and therefore you will work harder to achieve your ambition.'

8 | Elbow grease: Results day (and pre-results day)

You have done all of the hard work – your personal statement, the interview, the examinations – and you are now waiting for your results, the results that will determine whether you have achieved what you need to take your university place. This chapter explains what happens when you get your results and, if you have achieved grades or scores that are either better or worse than expected, what other options are available to you.

UCAS Extra

But what if you have not received any offers at all? If you find that you do not have any offers from your five choices, then you are eligible for UCAS Extra. This opens in February of each year and allows any student not holding an offer – or, any student willing to reject any offer they currently hold in favour of a new option – to apply for one additional university choice. If a university has spaces available left on a course, then they will advertise this on the UCAS website as an Extra place. This will not be available for all courses; however, it does help give students an extra – perhaps more prudently chosen – option before results days.

When the results are available

- A levels – second week of August
- IB – early in July
- Scottish Highers – first week of August

Ask your school or college for the exact date and time that they will issue you with the results.

Whichever of the exam systems you are sitting, you need to act quickly if you:

- have missed the grades or scores that you require to satisfy your firm offer

- are not holding any offers and wish to apply through Clearing (see below)
- wish to use the Adjustment system (see page 115).

If you have gained the grades that you need to satisfy your firm choice – congratulations, you have your place! The university will contact you with confirmation of the place.

What to do if things go wrong during the exams

Occasionally, students will underperform in an examination through no fault of their own. This could be through distressing family circumstances (a serious illness to a family member, for example), illness in the run-up to the exam (or during the exam) or unforeseen circumstances such as late arrival to the exam due to problems with public transport. In all cases, you should inform the universities that this has happened to you immediately after the examination. You should, if possible, get your referee to give the details to the universities and provide documentary evidence, such as a letter from your GP.

What to do if you have no offer

Students who are not holding any offers when the examination results are published (or those who have failed to achieve the grades that they need) can apply for vacancies through Clearing. The Clearing system operates by publishing all remaining university vacancies on the UCAS website and in the national newspapers. Students can then find appropriate courses and apply directly to the university. This time, you do not go through UCAS. UCAS will send you a Clearing Passport, which, when you have been made a verbal offer that you wish to accept, you then send to UCAS to confirm the place. Bear in mind that Clearing places at top universities are scarce, and so you will need to act very quickly.

You can also choose on your UCAS application form to allow universities to contact you with details of possible available places in advance of results day.

Clearing Plus

This is an innovation by UCAS introduced for 2020 which may be helpful to students who did not meet their firm or insurance offers.

- If a student finds themselves in Clearing, UCAS will personally match the student to courses they may be interested in, using what they know about them from their application, and what universities and colleges are looking for in a student.

- If a student is unplaced or has started a new application, a button will appear in Track – 'My matches'. This will take a student to their top 50 course matches. They will then be able to tell universities and colleges that they are 'interested' in their courses. If the university still has vacancies and the student meets their entry requirements, they may call the student.
- If a student wants to apply for something different than is on their list of matches then that is absolutely permitted – they can still find courses available in Clearing using the UCAS search tool (see above).

What to do if you have an offer but miss the grades

If you have gained grades that nearly meet those required for your firm choice (e.g. BBB for an ABB offer) but are good enough for your insurance offer, your first choice can still accept you. Otherwise, you are automatically accepted onto the insurance place. Check on Track to see if you have been accepted. If not, contact the university and see if it can be persuaded to accept you – see sample email below. Your referee might be able to help with this.

To: anthony.john@allington.ac.uk
From: Jonathan Luke
Subject: A level results

Dear Dr John
UCAS no. 08–123456–7

I have just received my A level results, which were:
Mathematics A*, Physics A, Chemistry C.

I hold a conditional offer from Allington of ABB and I realise that my grades may not meet that offer as although I gained a higher grade in mathematics, I dropped a grade in chemistry. Nevertheless I am still determined to study engineering and I hope you will be able to still consider me for a place.

My Director of Studies is emailing you a reference. Should you wish to contact him, his details are: Mr S. Buckley, tel: 0123 456 7891, email: s.buckley@edinburghhs.sch.uk.

Yours sincerely

Jonathan Luke

If your grades are well below those required for your firm choice but satisfy your insurance offer, you will be automatically accepted onto the insurance place. Check on Track to see if the insurance offer has been confirmed. If there seems to be a delay, contact the university.

If your grades are below those needed for the insurance offer, you are now eligible for Clearing (see above). Use the UCAS website and national newspapers to identify suitable courses from the published vacancies and contact the institutions by telephone.

If you have missing results (for example, an 'X' on your results slip rather than a grade) this probably means that there is an administrative error somewhere, for example a missing coursework mark. Contact your school or college examinations officer immediately to sort out the problem. Contact your firm and insurance choices and explain the situation to them and ask them to hold your place until the problem has been resolved.

If you have achieved grades that are not good enough to get you a Clearing place, you can be accepted onto Access, diploma or foundation places, and then progress to a degree course. Alternatively, you can resit your A levels and reapply next year. Discuss these options with your school and your parents. Don't make hasty decisions – ask the university to extend its deadline if necessary.

What to do if you meet your offer but you've changed your mind about the course

If your grades satisfy one of your offers but you have changed your mind about the course you want to study, you can be considered for Clearing courses if you withdraw from your firm/insurance place. Contact UCAS to withdraw from your original place. Use the UCAS website and national newspapers to identify suitable courses from the published vacancies and contact the institutions by telephone.

What to do if you exceed your expectations and offer requirements

Students who have achieved grades that are better than the predictions on which both their firm and insurance choices were based have the opportunity to enter Adjustment. If this happens to you and you decide you would like to apply to an institution that asks for higher grades, you can put your application on hold for a short time (a week) in order to see if any of these institutions would be willing to offer you a place. When you register for Adjustment you do not lose your original offer, so you are still able to accept this if you cannot find a course at another university.

Bear in mind that there are unlikely to be places available for Adjustment candidates on the most competitive courses, but it is still worth having a look if you are in this situation.

Retaking your A levels

If you have not achieved the grades that you needed for your chosen universities, and you do not want to take the available Clearing places, you could consider retaking one or more A levels. In the days when most examination boards offered January sittings, retaking might have meant studying for one term to boost the grade. The period from January to September could then be used to earn money, gain more work experience or travel the world. But, apart from international A levels, A level exams are now only available in June, and so retaking will involve studying for another year, so you need to be sure that your university aspirations are genuine enough to give you the motivation to add this extra year to your studies. As the A level system is now fully reformed, barring the last Phase Three legacy-subject examinations, you will need to retake the entire two-year qualification again and therefore plan to be able to do this in just a single year – you do not want a repeat of the examination if you are underprepared.

Speak to your teachers about the implications of retaking your exams. Some independent sixth-form colleges provide specialist advice and teaching for students. Interviews to discuss this are free and carry no obligation to enrol on a course, so it is worth taking the time to talk to their staff before you embark on A level retakes. Many further education colleges also offer retake courses, and some schools will allow students to return to resit subjects, either as external examination candidates or by repeating a year.

Reapplying

Universities are usually happy to consider students who are reapplying, either because they did not get the required grades first time around, or because they did not receive any offers of places. It is worth contacting the university to check whether this is the case. Some will have policies on grade requirements for retake candidates, while others might ask for evidence of any extenuating circumstances that may have affected the previous application.

Tips

- If there were extenuating circumstances that affected your application, include a brief mention of this in the personal statement ('I was disappointed not to have achieved the required grades, because my studies were affected by illness, but this has made me even more determined to become an engineer') but leave the details to the referee.
- If you are retaking, you can use the extra term or extra year to add weight to your application, for example by gaining more work experience, taking up a new subject, enrolling in evening classes that are relevant to your application, and furthering your reading.

> *Engineering without imagination sinks to a trade.*
>
> **Herbert Hoover, engineer**

9 | Building bridges: Fees and funding

The cost of studying at university comprises two elements: the tuition fees charged by the university, and costs associated with accommodation, subsistence, travel, books, entertainment and other living expenses. The UCAS website (www.ucas.com) has full details of fees and support arrangements. This chapter explains how the tuition fees are calculated, and provides information on sources of funding for, or other help with, living costs.

Fees for undergraduate courses

The tuition fees that you will have to pay for undergraduate courses will depend on where you live and where you intend to study. Universities are allowed to charge home undergraduate fees up to £9,250 per year, as part of the government's new Teaching Excellence Framework (TEF), which assesses universities and colleges on the quality of their teaching. Tuition fee loans will also increase to cover the higher fees. The fee cap for students studying in Wales remains at £9,000, while fees for students in Northern Ireland are a maximum of £9,250 per year for 2021 entry. The various exceptions are noted below.

For UK students, the fees do not have to be paid at the start of each year. You are effectively given a loan by the government that you repay through your income tax once your earnings, as of 2018, reach £26,575 a year for students in England and Wales, and £18,330 for students in Scotland and Northern Ireland. So, if you never reach this threshold, you would not have to repay the fees.

- Students living in England are required to pay a maximum of £9,250 per year if they are studying in England, Scotland or Northern Ireland, and up to £9,000 if they are studying in Wales.
- Students from Scotland who study at Scottish universities are not required to pay tuition fees. They will have to pay fees of up to £9,250 if they study in England or Northern Ireland, and up to £9,000 if they study in Wales.
- Students living in Northern Ireland will pay up to £4,395 if they attend a university in Northern Ireland, up to £9,250 if they study in England or Scotland, and up to £9,000 if they study in Wales.

- Students from Wales pay up to £9,000 if they study in Wales, or up to £9,250 if they study in England, Scotland or Northern Ireland.

EU students and International Students

As of 23 June 2020, the government announced that EU, EEA and Swiss nationals will not be eligible for home fees status in England, and Northern Ireland, Wales and Scotland went on to confirm the same. From August 2021, EU students will need to check the fees at respective institutions to find out what the relative cost will be, which will be variable from institutions.

This will have a big implication on Scottish universities as there will no longer be an over-saturation of students per place as in previous years because of the EU free tuition status. Therefore, it is a good time to apply if you are an EU student, but it will cost you more.

International students will pay higher fees, determined by each university (see page 107).

Support and information

England

There are maintenance loans of up to £12,010 in London and £9,203 outside of London per year available for students living in England dependent on household income. Further details can be found at www.gov.uk and www.officeforstudents.org.uk.

Scotland

Bursaries and loans of up to £7,750 a year are available from the Student Awards Agency for Scotland (www.saas.gov.uk) dependent on household income.

Wales

For 2021 entry, Student Finance Wales (www.studentfinancewales.co.uk) provides Welsh students with a fee grant and a fee loan of up to £12,260 for study in London, £9,810 for study outside of London, and £8,335 if you are studying while living at home. The balance between the fee loan and the fee grant is dependent on household income. Learning and special support grants are also available.

Northern Ireland

Student Finance NI (www.studentfinanceni.co.uk) provides loans up to £6,428 for study in London and £5,388 for study outside of London. Maintenance and special support grants are also available.

Postgraduate courses

Fees for postgraduate courses are determined by the individual universities, and will be different for home students and international students. For information about tuition fees, go to the university websites. Student loans for postgraduate courses are available from the Student Loans Company: www.slc.co.uk.

Sponsorship

Engineering students are more fortunate than their peers who are studying other subjects, because of the large number of sponsorship and bursary schemes available from engineering institutes, companies and the universities themselves. This is because it is recognised that the UK needs to attract more able students into engineering. The starting points for finding out about sponsorship are:

- the university engineering departments
- the engineering institutes (contact details are given in Chapter 11).

The engineering institutes or institutions are professional bodies that accredit and represent their members, provide training and information, promote their particular fields of engineering, organise or offer scholarships and help engineers with their careers.

The level of sponsorship varies from course to course, university to university, and institute to institute, and also changes from year to year. You will need to spend some time researching your options. The benefits of gaining sponsorship are numerous. In addition to financial assistance, you will have opportunities to gain internships, work experience or even a job to go to once you graduate. The very fact that an organisation or university has been prepared to sponsor you also says a good deal about your personal or academic qualities and will enhance your CV.

Sponsorship can include:

- financial aid during the degree course
- paid or part-funded work placements during holidays or a gap year
- work or study placements overseas
- improved chances of jobs with the sponsoring companies after graduation.

Several publications giving details of the scholarships and bursaries offered by educational trusts are available, including the *Directory of Grant Making Trusts*. You should also refer to the Educational Grants Advisory Service (www.egas-online.org.uk).

Scholarships

There are a variety of scholarships available for engineering students – possibly more than for any other discipline. The professional engineering institutes also offer some scholarships – for example, the Institution of Mechanical Engineers, mentioned below. You can also use the website www.scholarship-search.org.uk to look for undergraduate and postgraduate scholarships.

University scholarships broadly fall under three categories:

1.   scholarships offered by the university
2.   scholarships offered through donations by alumni
3.   industry scholarships, which can include work placements.

Normally, students apply for scholarships once they have accepted the offer of a place from the university, although sometimes universities will offer a scholarship alongside the offer of a place in order to attract the best students. There is a section on scholarships under the 'Fees and funding' heading on all university websites.

The Institution of Mechanical Engineers (IMechE)

Undergraduate scholarships and grants offer assistance to students who are about to start or have already started on mechanical engineering degrees accredited by the Institution. Current undergraduate scholarships are sponsored by Rolls-Royce, Land Rover and The Eaton to assist students with their accredited degree-level programmes. Annually, IMechE awards up to £800,000.

- IMechE Undergraduate Scholarships (£2,000 per year for up to four years)
- IMechE 'Land Rover Spen King Sustainability Award' (£2,000 per year for up to four years)
- IMechE Eaton Undergraduate Scholarship (£2,000 per year for up to four years)
- First year Undergraduate Scholarship (£9,000 in total)
- Whitworth Scholarship up to £20,000 to fund undergraduate degree
- Apprentice of the Year (£1,500)
- Overseas study award up to £2,000
- Group project award up to £500.

Source: www.imeche.org.

Case study

Peter is currently studying mechanical engineering at the University of Exeter. Wishing to join the Royal Electrical and Mechanical Engineers, he is being sponsored through his degree on the Army Undergraduate Bursary – Technical (AUBT).

'Having been in the cadets at school, I was sure that I wanted to join the army. Moreover, I have grown up with a family that loves cars and have been tinkering since a child, so engineering was always a likely path for me. My grandfather was an engineer during and after the Second World War, and I wanted to follow in his footsteps.

'With the high price of tuition fees nowadays, I needed funding support in order to go to university. I found the AUBT through my careers office at school and passed the formal assessments to get a grant of £16,000. This has helped me realise my goal and on graduation I will go into the service to pay back the investment they have made in me and to craft my own career.

'The Royal Electrical and Mechanical Engineers corps is deployed all over the world and their function is to keep the army moving. The variety of machinery I am learning to work with should prove invaluable after I leave the service as well. I enjoy working with all types of mechanical, electrical and aeronautical equipment to ensure they are operationally ready.

'I would encourage anyone to look closely at a grant or scheme if it helps them to study at university. The system is not particularly arduous, it just requires diligence at filling in the forms and an interview in some cases to ensure you are serious about what you are planning to embark upon, and the rewards vastly outweigh all of that.'

Did you know ... ?

Theoretical physicist Richard Feynman (1918–88) used to crack vaults for fun in his spare time.

10 | A cog in the machine: Further training, qualifications and careers

There is no 'typical' day for a working engineer. One of the attractions of engineering as a career is that you are, with suitable planning, able to follow a career path that suits your own individual skills and ambitions. If you like working outdoors, working as an on-site civil engineer would allow you to do this, whereas if you enjoy working in a laboratory then you might choose to be a structural or electronic engineer. As we saw in the introduction to this book, engineers can work as part of a large team for a multinational engineering company or on their own in their own company. Engineers who are interested in planning and finance can work as product engineers, assessing the economic viability of manufacturing a product and then designing the production line.

Some students with engineering degrees decide to change direction after graduating. For example, engineers are highly sought after in the financial sector because an engineering degree demonstrates that the student has analytical and problem-solving skills.

The engineering institutes (see list in Chapter 11) and university engineering departments provide a good starting point for further investigation of possible careers through the case histories that they publish. Almost all of the engineering organisations have sections on their websites (often under the 'Education' tab) that contain profiles of undergraduate and qualified engineers. Some good examples can be found:

- on the Royal Academy of Engineering's website: www.raeng.org.uk/education/this-is-engineering
- under 'Who are civil engineers' on the education pages of the Institution of Civil Engineers website (www.ice.org.uk/what-is-civil-engineering/who-are-civil-engineers)
- on the Try Engineering website (www.tryengineering.org)
- at www.whynotchemeng.com (for chemical engineering)
- on the Tomorrow's Engineers website (www.tomorrowsengineers.org.uk).

The websites of the university engineering departments are also useful sources of case histories, and they often have links that allow you to address questions to undergraduate engineers.

In its booklet *Educating Engineers for the 21st Century*, the Royal Academy of Engineering provides an overview of what qualities future engineers will need to possess:

> *No factor is more critical in underpinning the continuing health and vitality of any national economy than a strong supply of graduate engineers equipped with the understanding, attitudes and abilities necessary to apply their skills in business and other environments.*

> *Today, business environments increasingly require engineers who can design and deliver to customers not merely isolated products but complete solutions involving complex integrated systems. Increasingly they also demand the ability to work in globally dispersed teams across different time zones and cultures. The traditional disciplinary boundaries inherited from the 19th century are now being transgressed by new industries and disciplines, such as medical engineering and nanotechnology, which also involve the application of more recent engineering developments, most obviously the information and communication technologies. Meanwhile new products and services that would be impossible without the knowledge and skills of engineers – for instance the internet and mobile telephones – have become pervasive in our everyday life, especially for young people.*

> *Engineering businesses now seek engineers with abilities and attributes in two broad areas – technical understanding and enabling skills. The first of these comprises: a sound knowledge of disciplinary fundamentals; a strong grasp of mathematics; creativity and innovation; together with the ability to apply theory in practice. The second is the set of abilities that enable engineers to work effectively in a business environment: communication skills; team-working skills; and business awareness of the implications of engineering decisions and investments. It is this combination of understanding and skills that underpins the role that engineers now play in the business world, a role with three distinct, if interrelated, elements: that of the technical specialist imbued with expert knowledge; that of the integrator able to operate across boundaries in complex environments; and that of the change agent providing the creativity, innovation and leadership necessary to meet new challenges.*

> *Engineering today is characterised by both a rapidly increasing diversity of the demands made on engineers in their professional lives and the ubiquity of the products and services they provide. Yet there is a growing concern that in*

*the UK the education system responsible for producing new
generations of engineers is failing to keep pace with the inherent
dynamism of this situation and indeed with the increasing need
for engineers.*

Source: www.raeng.org.uk/publications/reports/educating-engineers-
21st-century
Reproduced courtesy of the Royal Academy of Engineering.

Chartered Engineer status

Chartered Engineer (CEng) is a professional title registered by the
Engineering Council. Engineers who achieve this status have been
able to demonstrate that they have reached a high level of professional
competence. Attaining the status of Chartered Engineer brings many
benefits, including:

- being part of an elite group of highly qualified engineers
- professional recognition of your qualifications and attainments
- higher earnings potential
- improved career prospects
- international recognition of your academic and professional qualifications
- access to continuing professional training.

As shown in Figure 4 on the following page, the normal route towards
gaining the qualification is:

1. an accredited bachelor's degree (BEng)
2. an undergraduate master's degree (MEng)
3. membership of one of the professional engineering institutes
4. experience of professional practice.

It usually takes between eight and 12 years from the start of an under-
graduate degree to reach CEng status. For more details, contact the
Engineering Council or one of the engineering institutes (see the list in
Chapter 11).

Incorporated Engineer status

While this book focuses on you obtaining Chartered Engineering status
(CEng), there is also Incorporated Engineer status (IEng), which we will
mention here in brief.

The fundamentals of engineering form the base of both qualifications;
however, there are key differences. The most common difference is

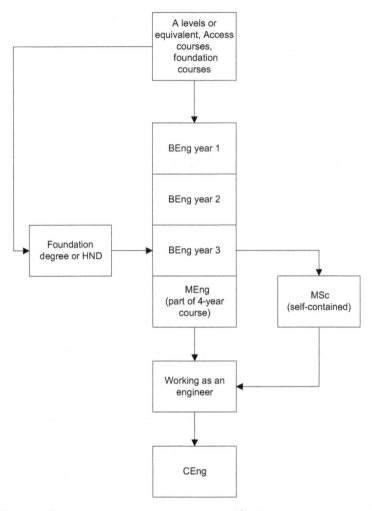

Figure 4 Routes to gaining an engineering qualification

that Chartered Engineers are considered to work towards developing solutions to engineering problems, whether that be in the creation or in the utilisation of new technologies. Incorporated Engineers are then responsible for the maintenance of applications and management.

The route that you embark on is generally determined by your employer and dependent on the skills and competencies they will require of you. Therefore, you should be careful when choosing your graduate scheme that you know what you are working towards.

If you wish to become an IEng, generally you would take a Bachelors degree first, though it is possible to become accredited in this qualification with a foundation degree or HND and then further learning.

You do not need to become an IEng before becoming a CEng. Although gaining a CEng may bring better pay compared with an IEng, you should choose your route based on what you are interested in rather than on financial reward.

Master's courses

Many undergraduate engineering courses are four years in length, and lead to a master's qualification (MEng) rather than a bachelor's degree (BEng). The alternative route to a postgraduate qualification is by taking a self-contained MSc course after completing the bachelor's degree. Master's degrees allow students to focus their studies on a specific area of engineering. Applications for self-contained postgraduate courses are usually made directly to the universities, rather than through a central scheme. For listings of master's courses, see www.prospects. ac.uk and the university websites.

The advantages of doing a master's course are:

- greater specialisation
- better job prospects
- higher earning potential.

The cost of a master's course varies from university to university, but is likely to be in the range of £7,500–£16,000 per year for tuition fees. A number of grants and scholarships are available (see pages 119–20).

The structure of a master's course will depend on where and what you study. As an example, here is the current course programme for the MSc Energy Engineering with Environmental Management at the University of East Anglia:

- Electrical energy generation, distribution and storage
- Energy engineering dissertation
- Energy engineering fundamentals
- Oil and gas engineering
- Wind energy engineering
- Options: including nuclear and solar energy; energy futures; team energy project; wave, tidal and hybrid energy engineering.

Source: www.uea.ac.uk/course/postgraduate/msc-energy-engineering-with-environmental-management.
Reprinted with the kind permission of the University of East Anglia.

Career opportunities and employment prospects

The engineering industry contributes around 25% of the UK's economic turnover (precisely 23.2% of the UK's £5.3 trillion turnover from registered enterprises, as quoted in *Engineering UK 2018*), and employs nearly 6 million people in over half a million businesses. Employment prospects for UK engineering graduates are good, and the manufacturing and engineering sectors look set to grow significantly. The organisation Engineering UK estimates that the UK will need over two million more engineers within the next five to 10 years. According to Prospects.ac.uk, by 2024 a further 186,000 people with engineering skills will be needed per year.

According to figures produced by the Higher Education Careers Service Unit (HECSU) and the Association of Graduate Careers Advisory Services (AGCAS) for *What Do Graduates Do*, employment rates for recent engineering graduates were higher than the average for all graduating students, and the unemployment rates lower. For example, around 79% of graduating civil engineering graduates found employment within six months of graduating, with a further 13% embarking on further study or combining work and study. These figures compare with the averages for all graduates of 76% employment and 6% unemployment. The report also highlights the attractiveness of engineering graduates to other employers, with significant numbers finding employment within the managerial, IT and financial sectors.

Source: www.hecsu.ac.uk.

In its report on employment, Engineering UK highlights the higher-than-average starting salaries for engineering graduates. With a mean starting salary of over £25,000 (£36,500 roughly for chemical and medical engineers), engineering was the fourth highest sector and around 16% higher than the average for all graduates. As a comparison, the average starting salary for the art and design sector was £16,000. The average working salary for an engineer as their career progresses is £40,000, with the top 10% of senior engineers earning upwards of £110,000.

Source: www.engineeringuk.com.

> **Did you know ... ?**
>
> The Empire State Building in New York City is actually designed to be a lightning rod and is struck about 100 times every year.

11 | Light years ahead: Further information

Useful contacts

University applications

www.ucas.com

Funding

www.gov.uk
www.slc.co.uk
www.saas.gov.uk (Scotland)
www.studentfinancewales.co.uk (Wales)
www.studentfinanceni.co.uk (Northern Ireland)

News

www.bbc.co.uk/news
www.theguardian.com

Overseas voluntary projects relevant to engineering

www.ewb-uk.org/the-work

Organisations for engineers in the UK

Engineering Council
www.engc.org.uk

Institution of Civil Engineers (ICE)
www.ice.org.uk

Institution of Chemical Engineers (IChemE)
www.icheme.org

Institution of Mechanical Engineers (IMechE)
www.imeche.org

Institute of Electrical and Electronics Engineers (IEEE)
www.ieee-ukandireland.org

Institution of Structural Engineers (IStructE)
www.istructe.org

Royal Academy of Engineering
www.raeng.org.uk

Chartered Institute for IT
www.bcs.org

Chartered Institution of Building Services Engineers (CIBSE)
www.cibse.org

Chartered Institution of Highways and Transportation (CIHT)
www.ciht.org.uk

Chartered Institute of Plumbing and Heating Engineering (CIPHE)
www.ciphe.org.uk

Chartered Institution of Water and Environmental Management
(CIWEM)
www.ciwem.org

Institute of Acoustics (IOA)
www.ioa.org.uk

Institute of Cast Metals Engineers (ICME)
www.icme.org.uk

Institute of Highway Engineers (IHE)
www.theihe.org

Institute of Marine Engineering, Science and Technology (IMarEST)
www.imarest.org

Institute of Measurement and Control (InstMC)
www.instmc.org.uk

Institute of Materials, Minerals and Mining (IoM3)
www.iom3.org

Institute of Physics (IOP)
www.iop.org

Institute of Physics and Engineering in Medicine (IPEM)
www.ipem.ac.uk

Institute of Water (IWO)
www.instituteofwater.org.uk

Institution of Agricultural Engineers (IAgrE)
www.iagre.org

Institution of Diesel and Gas Turbine Engineers (IDGTE)
www.idgte.org

Institution of Engineering Designers (IED)
www.institution-engineering-designers.org.uk

Institution of Engineering and Technology (IET)
www.theiet.org

Institution of Fire Engineers (IFE)
www.ife.org.uk

Institution of Gas Engineers and Managers (IGEM)
www.igem.org.uk

Institution of Royal Engineers (InstRE)
www.instre.org

Nuclear Institute (NI)
www.nuclearinst.com

Royal Aeronautical Society (RAeS)
www.aerosociety.com

Royal Institution of Naval Architects (RINA)
www.rina.org.uk

Society of Environmental Engineers (SEE)
https://socenv.org.uk/page/see-support

Society of Operations Engineers (SOE)
www.soe.org.uk

The Welding Institute (TWI)
www.theweldinginstitute.com

Specialist publications

Architect magazine
www.architectmagazine.com

The Engineer
www.theengineer.co.uk

Aviation Week
www.aviationweek.com

Nano
www.nanomagazine.co.uk

Race Car Engineering
www.racecar-engineering.com

Engineering News Record
www.enr.com

New Civil Engineer
www.nce.co.uk

E&T [Engineering and Technology] magazine
https://eandt.theiet.org

Electronic Engineering Times
www.eetimes.com

Structure magazine
www.structuremag.org

Books

Engineering

- Blockley, David, *Bridges: The Science and Art of the World's Most Inspiring Structures*, OUP, 2010.
- Brain, Marshall, *The Engineering Book: From the Catapult to the Curiosity Rover*, Sterling, 2015.
- Brenner, Brian, ed., *Don't Throw This Away!: The Civil Engineering Life*, American Society of Civil Engineers, 2006.
- Browne, John, *Seven Elements That Have Changed The World: Iron, Carbon, Gold, Silver, Uranium, Titanium, Silicon*, Weidenfeld and Nicholson, 2013.
- Dupre, Judith, *Skyscrapers: A History of the World's Most Extra-ordinary Buildings*, Black Dog & Leventhal Publishers Inc., 2008.
- Dyson, James, *Against the Odds: An Autobiography*, Orion, 1997.
- Eberhart, Mark E., *Why Things Break: Understanding the World by the Way it Comes Apart*, Three Rivers Press, 2004.
- Fawcett, Bill, *It Looked Good on Paper: Bizarre Inventions, Design Disasters and Engineering Follies*, Harper Paperbacks, 2009.
- Gordon, J.E., *Structures: Or Why Things Don't Fall Down*, DaCapo Press, 2003.
- Gordon, J.E., *The New Science of Strong Materials: Or Why You Don't Fall Through The Floor*, Penguin, 1991.
- Hart-Davis, Adam, *Engineers*, Dorling Kindersley, 2012.
- Michell, Tony, *Samsung Electronics and the Struggle for Leadership of the Electronics Industry*, John Wiley & Sons, 2010.
- Monroe, Randall, *What If: Serious Scientific Answers to Absurd Hypothetical Questions*, John Murray, 2015.
- Munroe, Randall, *Thing Explainer: Complicated Stuff in Simple Words*, John Murray, 2015.
- Open University (author), *Engineering: the Nature of Problems*, Open University, 2016.
- Petroski, Henry, *Invention by Design: How Engineers Get From Thought to Thing*, Harvard University Press, 1998.

- Sammartino McPherson, Stephanie, *Tim Berners-Lee: Inventor of the World Wide Web*, Twenty-first Century Books, 2009.

Thinking skills

- Butterworth, John and Thwaites, Geoff, *Thinking Skills*, CUP, 2005.
- Tanna, Minesh, *Think You Can Think?*, Oxbridge Applications, 2011.

Joke

Customer: Do you have any two-watt, four-volt bulbs?
Sales Rep: For what?
Customer: No, two.
Sales Rep: Two what?
Customer: Yes.
Sales Rep: No.

Glossary

UCAS applications

Adjustment
Runs in August and allows students who have performed better than expected to look at upgrading their university places.

Admissions tutor
Someone within an engineering department who deals with UCAS applications.

Clearing
The period from early July to mid-September when students who are not holding offers for undergraduate university places can approach universities that still have vacancies.

Deferred entry
Applications for an undergraduate place for the following year, allowing the student to take a gap year.

Extra
Allows students who are not holding any offers to approach extra universities, prior to receiving their examination results.

Gap year
A year between leaving school or college and starting university, usually used to gain further work or life experience or extra qualifications.

IELTS
The International English Language Testing System. Students who do not have English as their first language must reach a certain IELTS level in order to gain entry to study in the UK.

Legacy and new specification A levels
From September 2015, new A level specifications, in which the A level is examined at the end of the course rather than combining AS and A2 scores, were introduced and are being phased in over a four-year period. The full A level examinations for the first group of reformed subjects were sat in June 2017.

UCAS
The Universities and Colleges Admissions Service, the online undergraduate application system.

Engineering

Aerospace engineering
A specialist branch of mechanical engineering focusing on aviation.

Automotive engineering
A specialist branch of mechanical engineering dealing with transport.

BEng
The qualification gained after a three-year undergraduate engineering degree course (often four years in Scotland).

Biomedical engineering
Linking engineering and living things.

Chemical engineering
The branch of engineering that looks at industrial processes involving chemicals, drugs, food and fuels.

Civil engineering
The branch of engineering dealing with large-scale infrastructure projects such as roads, bridges and dams.

Electrical engineering
Engineering involving electrical devices. Often taught as a joint degree with electronic engineering.

Electronic engineering
The branch of engineering that deals with electronics, such as integrated circuits.

Energy engineering
Engineering associated with energy production and transmission, environmental issues and alternative energy sources.

Mechanical engineering
The branch of engineering dealing with machinery.

MEng
The qualification gained from a four-year engineering degree course (often five years in Scotland).

MSc
A self-contained master's postgraduate degree.

NRA
National Risk Assessment.

Production engineering
Engineering that relates to manufacturing processes.

Structural engineering
Engineering that deals with the use of suitable materials for engineering projects.

EDUCAD

EDUCATION
CONSULTING